T

Alarm Book

The
Alarm Book

A Guide to Burglar and
Fire Alarms

Dan McTague
Doug Smith

Butterworth-Heinemann
Boston   London   Oxford   Singapore   Sydney   Toronto   Wellington

Library of Congress Catalog Card Number 91–76607
ISBN 0–7506–9316–9

Butterworth-Heinemann
80 Montvale Avenue
Stoneham, MA 02180

10 9 8 7 6 5 4 3 2 1

Printed in the United States of America

Dedicated to Robert Hannum

Authors' Note:

THE ALARM BOOK is the result of a collaboration which began several years ago out of our desire to provide a practical and objective handbook on the professional alarm industry. Dan McTague contributed seven years' experience engineering and marketing commercial alarm systems. Veteran news reporter Doug Smith directed the writing to keep it objective and informative to the nontechnical reader.

Acknowledgement:

The authors would like to extend their appreciation to numerous friends in the alarm industry for their cooperation in this project. We owe special thanks to API Alarm Systems, Los Angeles, for supporting this book in its formative stages. We also thank API and Alarm Device Manufacturing Co (Ademco), New York, for providing photographs from their libraries. Finally, we are indebted to Ron Dalton of API for his technical comments on the final manuscript.

Notes on Style:

Throughout this book, we frequently use the term UL Approved. Actually, Underwriters Laboratories of Santa Clara uses a different term, UL Listed, for products and services which it has determined meet its performance specifications. As a matter of style, we have adopted the popular usage, UL Approved, because we think it more accurately represents the accepted force of UL Listing in the marketplace. Underwriters Laboratories does not discourage this usage.

As much as possible, we avoid the use of trade names to describe the products and services of professional alarm companies. It is common in the alarm industry, however, to use trade names rather than generic names to describe the products of a particular alarm company. On occassion trade names have slipped into our text. This sould not be considered an endorsement of one product over another, as there are usually many products that do the same task effectively. You may find an alarm company using a product that does not appear by name in this book. That does not mean it is necessarily of lesser quality than a similar device we name. All photos used in this book are for the purpose of example and do not constitute a recommendation of a particular manufacturer's product.

TABLE OF CONTENTS

Part I
UNDERSTANDING ALARMS

THE TOOLS OF SECURITY   41

FIRE, HAZARD AND WATER FLOW 97

HOME ALARMS 117

Part II
LIVING WITH ALARMS

BUYING THE ALARM 139

WORKING WITH ALARMS 157

THE NIGHTMARE ALARM 167

Appendices

Part I
UNDERSTANDING ALARMS

GETTING IN FOCUS

When you purchase an electronic alarm system, you are making an investment. Like any other investment, your alarm system can yield a return or it can waste your money, depending on the decisions you make.

When compared to the relatively small investment alarm systems require, their positive potential is enormous. A successful alarm system can yield years of uninterrupted, prosperous operation of your business, all in an atmosphere in which you and your employees feel secure. At home an alarm system can extend that same security to you, your family and your possessions both day and night and whether you are in your home or elsewhere.

Beyond that, there is a more abstract—but still real—benefit, the power that comes to you from confronting a threat or danger you would probably prefer not to think about at all and knowing that you have done all you can to counter it.

On the other hand, an unsuccessful security system will surely lead to aggravation, worry and exposure to financial setback or physical harm to you, your employees and your family.

Creating good security is, like most things, difficult when you don't know how and easy when you do. Unfortunately, the purchase of an alarm system is too often viewed not as an investment, but simply as an unwanted expenditure and a burden. And too often the businessman, manager, or homeowner is thrown into the arena of security with inadequate knowledge and experience, making decisions difficult.

If you are like most people, your interest in the nuts and bolts of electronics is nil and you are investigating this book out of a sense of need rather than curiosity, probably for one of the following reasons:

This book is for you if • • • • •

- Your business or home has been the victim of robbery, burglary or fire.

- You are concerned about that possibility.

- Your insurance carrier has asked you to install an alarm system to qualify for insurance.

- You are an insurance agent and you must work hand-in-hand with customers on the approval of alarm systems.

- You already have an alarm system and it is giving you problems.

- You have just been assigned to handling security for your company.

We believe you will find security an interesting subject...

Our goal in this book is to give you the information you need to make your decisions easier and your alarm system more effective.

Real understanding, though, requires more than just information. It comes only with experience. So we have attempted to communicate the experience of working with a modern security system to help you get a grasp on good judgement quickly, so you

You will feel more in control when you know something about it...

will not regret decisions you make now in the light of future events.

We won't be over technical with electronics jargon and we do not plan to tell you how to build and operate your own alarm system. Experience tells us that where significant value is involved the best results in security are achieved by a professional alarm installation and service company.

And you WILL be more in control . . .

When you have a well designed alarm system installed.

Why then, if you are leaving the responsibility to a professional, should you read a book on security? The answer is that if reading this handbook allows you to avoid just one of the time consuming, costly and sometimes disastrous pitfalls of dealing with alarms and alarm companies, your time will be well rewarded.

When purchasing an alarm, you will run headlong into many decisions that you could hardly have foreseen. Which of the hundreds of companies advertising alarm services can you trust to do the job properly? How much money should you expect to spend? Where should you draw the line on adding exotic—and progressively more expensive—electronic devices? And—a question which may not seem very important to you until after your alarm system is installed and functioning—what do you need to do to make the system work as efficiently as possible?

We can assure you that as you gain experience with alarm systems, you will find this last question to be the one which plays the greatest role in achieving security. To explain this, we should make a distinction between security and alarms, two concepts which are often confused. Alarms represent the physical aspect of security, what we will call the tools of security. But security itself is a condition held and maintained by the constant will and attention of management. And this applies equally to the professional management of a business or the more casual management of a home. Security is like safety; you can't acquire it in a package. It develops through a continuing state of mind.

The alarm company and you...

An alarm system presents its user with a continuous management responsibility that is actually more complex than might initially be appreciated. Unlike an electrical or plumbing system that, once installed, tends to operate trouble free, an alarm system requires daily attention and occasional servicing.

A complex management relationship...

When dealing with most alarm companies, you will actually find yourself engaging that company on four different levels: A) engineering and installation of the system, B) ongoing monitoring by a central receiving station, C) maintenance of equipment and service response to alarms and D) management and accounting. With some companies, you may contract for the additional service of private armed response. Generally, the level of security achieved is directly related to the clarity of the management pattern set up by the alarm user to relate with the alarm company.

But remember...

For your part in this, there will be procedures to learn for turning the system on and off at the proper times. There will be periods of experimentation while sensitive equipment is being adapted to your property. And there will undoubtedly be late night telephone calls to disturb you. Your willingness to live with these annoyances is all a part of maintaining that condition we call security.

The only real security you have is that which you give yourself.

In short, the only security you have is that which you give yourself through the right use of the professional services that are offered.

An important aspect of achieving security is understanding the limits of alarm equipment. It would be easy to assume that today's sophisticated electronic surveillance techniques can offer some kind of absolute protection. In fact, this isn't the case at all.

The essential function of electronic alarms...

While there is a certain amount of deterrent value just in the fact that a property is protected by a professionally installed alarm, there is also a working axiom in the alarm industry that says, 'Alarms do not prevent burglary; they provide notification.' That notification is designed to protect your property by calling a response into action. Your security rests primarily upon the sureness of the notification and the speed of the response. Yet neither of these is guaranteed. Let us explain.

...is notification.

First, because alarm buyers are seldom willing to pay for the ultimate in high security equipment, there is no absolute guarantee that notification will occur every time there is an intrusion.

Second, when notification does occur, the response always relies on the actions of people. Electronic gadgetry is only one arm of an arsenal of protection that includes your local police agency, the employees of your alarm company, and, of course, yourself. It is the coordinated actions of all of these that makes security work.

Be advised...

There are no guarantees.

With these limitations in mind, we will begin our discussion of security with a brief theoretical overview of how security systems work. Here and throughout the book, security will be discussed primarily in terms of burglary in commercial applications. This is to simplify reading, not to limit the application of information to that context. In most cases the same information is equally applicable to other types of alarms, such as home or fire alarms, which will be discussed in separate chapters. Wherever distinctions are important, they will be clarified.

It is our hope that our book will prove useful as a reference on alarm decisions long after the first reading. For that reason, we occasionally repeat some important points to be sure you will find them when looking back through the book for the answer to a specific question.

Now, to help clarify the way alarms work, we will begin with a short analogy. We call it...

The Cycles of Security

You may not have thought of it this way, but every time there is a threat to security in your office or home, two distinct cycles of activity go into play. The success of your security effort depends on which cycle draws to its conclusion first.

The first, the **Event Cycle**, is touched off by an intruder or potentially hazardous event in your building. The second, the **Response Cycle**, is set off by the electronic devices which your alarm company installed there.

Cycles of Security

The event cycle begins when a burglar forces his way into a building you own or insure. Time ticks away as he first makes his entry, then searches for what he wants, gathers it into a central place, carries it out, and finally drives away. Unless interrupted, this cycle will result in a significant loss for you.

Response Cycle

 Alarm Signal

 Central Station

 Police Notification

Response

The goal of a modern silent alarm service company is to prevent that loss by making sure the original entry is detected, then by initiating a response cycle to interrupt the threatening event cycle before it runs its course.

Sometimes, careful installation of alarm equipment can stretch out the event cycle, forcing an intruder to leave himself at risk longer. One way, for example, is by making sure anti-burglary bars are placed inside the protection system. This forces a burglar to activate the alarm before attempting to penetrate the bars and may discourage him from trying.

Once detection occurs—and your intruder may very well know the exact moment it happens—the second cycle (the response cycle) begins. The level of protection achieved by alarm service is as good as the speed of its response cycle and no better.

In spite of the wonders of electronics, there is nothing automatic about this response. It depends on a complex set of occurrences, all subject to fortune, failure, and human error. To appreciate just how complex this is, try to picture the alarm system not as a single unit, but five component systems all working together.

Blueprint of An Alarm

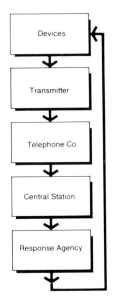

Blueprint of an Alarm

The **first** component system is installed in your building. It consists of electronic trip-points situated so they will be disturbed by any burglary attempt.

All of these trip-points are electrically connected inside your building to the **second** component system, the alarm transmitting devices. When one of the trip-points is triggered, a transmitter sends out a message, 'ALARM.'

This message is relayed by the transmitter over the **third** component system, leased telephone lines which can consist of anything from a private, direct line to the standard telephone lines and relays used to carry your voice when you make a call.

In the **fourth**, and most complex, component—an alarm company central station—the signal is received and analyzed to determine what action is needed.

Operators at the central station call the **fifth** and final component system into action by relaying the alarm, if necessary, for private armed response or response by the appropriate public agency.

We have, therefore, an alarm system, a transmitting device, a leased telephone connecting circuit, an alarm company central station, and a responding enforcement agency.

Because the alarm system is on their premises and visible to them, there is a tendency among subscribers to think of it as the most important element. It is our view, however, that of the five component systems the central station is the one in which the quality of an alarm company's service is most likely to be established.

While a failure in any part of an alarm system can negate its purpose, the central station is the one component which carries responsibility for maintaining the integrity of all the parts. It is the job of the central station to coordinate all the other components, to make final decisions involving all alarm activity, to oversee the maintenance of alarm equipment and to service not just one, but thousands, of separate alarm systems.

For these reasons, we will begin our discussion of alarm service with the central station. We will discuss how it works and what makes one better than another.

Before we take up the subject of central stations, let us end our introduction to security with a final point to help you focus your thinking on effective security.

False Alarms

...a corrosive influence on alarms

Outside of the burglar, whose presence is the reason for the burglar alarm industry, the major nemesis to be faced and overcome by everyone associated with an alarm system is the occurrence of false alarms. Easily 98% of all alarm signals received and responded to turn out to be false. The wasted energy and loss of confidence resulting from these false alarms defy imagination. A few of the consequences are: tremendous losses of manpower and money; the levying of fines by municipal agencies upon alarm companies and subscribers; the levying of false alarm charges by alarm companies upon subscribers; a tendency for alarm signals to be ignored by everyone involved; the deterioration of the relationship between police, alarm companies and subscribers, a general miasma throughout the industry.

Without belaboring the point any more here, let us just add that it is impossible to overstate the value of taking this problem into account in every decision you make. Prevention of false alarms is, therefore, a major consideration throughout this book.

Inside A Modern Central Station

Courtesy American Protection Industries, Los Angeles

THE CENTRAL STATION

Our first step in appraising the central station (or stations) of any company is simple: either it is approved by Underwriters Laboratories (UL) or it is not.

This is the same UL whose initials you see on the stickers attached to electrical appliances used in the home as well as a wide variety of products throughout America.

UL is an independent, nonprofit testing laboratory created many years ago out of the need of the insurance industry to subject merchandise and services to a uniform standard of safety and reliability. UL defines the standards and conducts tests to see that the standards are met.

While not mandatory, the Underwriter Laboratories seal has come to be recognized as the universal standard of excellence in many fields, including the burglar alarm industry.

Unlike many service industries which operate under strict governmental regulation, the alarm industry (outside of insurance underwriters' requirements) has been left to set its own standards and to regulate and police itself. Furthermore, though it has the day-to-day responsibility to protect human life and billions of dollars of property, the alarm industry has no regulatory body of its own with any real power comparable to that of the American Medical Association or the Bar Association.

In the vacuum created by the absence of government or self regulation, the insurance industry has provided several fine organizations for rating and approving various alarm methods. However, Underwriters Laboratories provides the universally agreed upon standards of performance.

What Does 'Non-Approved' Mean?

Approval by UL is not a requirement for operation of a central station alarm company. It is an option. Prior to the latest developments of technology, UL Approval had been a costly and time-consuming investment for an alarm company and was often not sought until the number of the company's subscribers grew to the point that the additional investment was feasible. The construction of a UL Approved central station today, however, is a much simplified matter. And, as electronics have improved, the cost has come within reach of many smaller alarm companies.

There are good alarm companies that are not UL Approved...

There are some circumstances under which UL Approval is not generally considered necessary to the successful operation of a central station alarm company. This would be true in rural areas and communities where the crime rate is relatively low. In such areas, UL Approved central stations are often not even available and insurance companies do not require the certification of alarm systems by UL.

...where integrity is the rule.

Even in urban areas, many non-approved central stations give service that is consistent with UL Standards, though they are not UL Approved. These companies instead rely upon their own integrity and understanding of alarm service. Generally they will perform at a level consistent with UL Approval.

If you know someone in a business similar to yours who has had a good relationship with an alarm company, UL Approved or not, that company merits your consideration. Conversely, a fully approved company with a bad reputation should be scrutinized carefully because integrity can't be legislated. The recommendation of a satisfied customer is as important in the alarm industry as in any other, none of which takes away from objective standards such as those of UL.

However, the burden of judgement falls upon you.

Without UL Approval, there would be no standard for comparison. It would be possible for anyone to create an alarm company merely by wiring his customers' alarm trip-points into a signal panel in his bedroom at home. Whenever an alarm would go off it would wake him up...he'd phone the police... and go back to sleep. But he could still advertise his service as being a 'central station silent alarm' company!

Not many companies work quite so crudely, but an inspection of all non-approved companies that purport to have 'silent alarm service monitored by a central station' would reveal methods and standards of operation that range from none at all to better than UL Approved.

However, unless you have the knowledge of an exceptionally complex field to make your own judgement of an alarm company's performance, UL Approval is virtually the only independent guideline you have.

Note: As will be discussed in a later section, fire alarms are strictly regulated by the National Fire Protection Association (NFPA) Standards. UL follows these Standards in its approval of central stations.

What Does 'UL APPROVAL' Mean?

'UL Approved' means that the central station is constructed and operated according to the rules and standards set by Underwriters Laboratories, Inc.

A few of the requirements of a UL Approved central station are:

UL ensures • • • • •

- a fire and bomb resistant building with its own standby power (in case of community power failure) and a security-controlled structure with all construction and equipment conforming to UL Certification Standards.

- UL Approved alarm receiving equipment.

- adherence to the rules set by UL in all functions from equipment operation and paperwork to response criteria.

- a permanent time-stamped record of all activity relating to each approved account.

- on-file instructions describing the special handling of alarm signals for each account.

- an established number of personnel present at all times (determined by the time of day, etc.) including: a supervisor in charge, attendants at the signal receiving board, runners to respond to alarms (within the minimum time allotments for UL Certified systems), adequate service personnel to assure a continuous function of alarm systems.

...and more...

To list all UL requirements in the UL Standards would be impractical here, since these fill many volumes of UL Specifications Manuals. A more detailed summary is given in Appendix A at the end of this handbook.

...close to 100 years of developed standards.

If you require a company that is UL Approved, be aware that most are not. In the Greater Los Angeles area, for example, there are more than 200 burglar alarm companies, yet only about a dozen of these have central stations that are UL Approved.

Better Than 'UL Approved'

While on the subject of UL Approval, we should point out that there are no rules against any alarm company exceeding UL requirements.

Some circumstances call for more than UL requires. Any alarm company that conscientiously tries to give its customers the 'best' protection there is will certainly offer many services that are not required by UL. Some highly valuable services offered by the better alarm companies but not required by UL are:

Such as • • • • •
- radio-dispatched patrol vehicles manned by armed patrolmen for alarm response and protection of burglarized or endangered premises in emergencies.

- 24-hour availability of service supervisors to oversee restoration of alarm systems at any hour.

- permanently maintained inventory of all alarm equipment parts so that any part or any device is always immediately available to restore service.

Changing Technology In The Central Station

...an industry in transition One function of the modern central station that may not be readily apparent is the rapid storage and retrieval of important information on thousands of subscriber accounts.

To provide the shortest possible response cycle whenever an alarm is received, or to know what to do when a customer calls in for a routine check or service, the central station must have quick access to detailed records for that particular user's alarm system, including codewords, procedures, type and number of trip- points installed, a history of prior alarm activity and record of service calls.

Computer technology is...

The ability of modern computer technology to cross-reference such records and display them instantly before the central station operator can greatly increase the efficiency of alarm response and service. But it should not be assumed for this reason that this technology is in standard use throughout the alarm industry.

Though UL specifies that records of all alarm activity must be kept, UL Standards can still be met by a company that runs its central station without the advantages of modern computer technology.

In the alarm industry today, a great gap exists between the archaic procedures and equipment still used by some companies and the advanced technologies applied by others in the struggle to keep abreast of rapidly growing demands.

...simplifying decisions...

Some central stations, for example, can be found still using signal switching devices which are the equivalent technologically of telephone equipment installed at the end of World War II. In such a scene, large banks of mechanically clicking relays are receiving and sorting signals coming from hundreds of separate buildings. The electrical impulses are translated into a hashmark code printed out by ticker-tape machines. Interspersed with the ticker-tape machines is other equipment blinking and beeping with information from other types of alarm systems. In addition to interpreting the beeping and blinking signals, the operator must decode the stream of hashmarks into a numerical sequence and then match that to a specific subscriber.

...speeding response... These methods have proved effective over many years of service. But, while fairly reliable for the volume of work originally intended, they cannot be expected to handle the expanding message load of the growing alarm service market.

As might be suspected, this type of operation is very vulnerable to human error. During busy periods, the operators work under tremendous pressure, performing multiple operations and translating alarm signals from codes to identified addresses. Mistakes of judgement often send police hurrying to an address where nothing unusual is happening while, somewhere else, a burglar works unhindered.

Reading Alarm Hashmarks

...reducing human error

To the general improvement of the industry, computer technology is replacing the ticker-tape era. In the earliest jump of technology, solid state electronics eliminated the mechanical relays and the multitude of problems which went with them. Still newer equipment came along to translate the incoming signals to printed messages at electronic speed. The operator then only had to interpret a printed code. In the most up-to-date central stations, signal interpretation is no longer even necessary. Signals from your business and others go directly from the receiving equipment to a computer which automatically performs most functions. Only signals which demand a special response are brought to the attention of central station operators, and this is done in written English.

The importance of innovations that eliminate the potential for human error can more easily be appreciated as we describe just what happens inside the central station when an alarm goes off.

What Is The Response To An Alarm?

Five steps occur...

When the central station receives a signal via an alarm transmitter at a protected site saying that one of its trip-points has been disturbed, the central station reacts with a specific predetermined course of action.

It must:

One:
A reference check

Check the signal against account instructions. For every subscriber there is an instruction card which helps the operator know what to do. It will tell him, for example, which alarm signals are anticipated... as when a shopowner opens in the morning or closes at night. This check can now be done automatically by a central station computer, leaving the attendant to handle only those signals which are not routine.

Two:
A pause to reflect

Verify the nature of the alarm. The central station operator will usually wait a minute or two after receiving a burglar alarm signal to allow a subscriber who has set an alarm off accidentally to call the central station (with the exception of holdup or fire). Through the use of a codeword the central station is assured that an authorized person is calling. By prearrangement, the central station operator will often initiate the call to the subscriber rather than wait. (In the case of home alarms this is almost always done.) If verification that the alarm is accidental is not forthcoming, it is then assumed that the alarm is real and the most critical step is taken.

Three:
A critical decision

Report the alarm. The central station operator phones the local police dispatcher in the case of burglary or sends a direct signal to the fire department in the case of fire. (Strict adherence to UL Standards would require this step to follow immediately). We would be less than candid, however, to suggest that this happens in every case. There may be times—as when a subscriber has a nagging habit of setting off his own alarm or when an alarm trip-point has a known propensity to misfire—that the central station operator will make a decision not to alert police immediately. He may decide to first send an alarm company patrol vehicle to investigate and alert the police later only if the need is observed. The inherent delay in this practice may not, in the long run, prove as costly as the loss of confidence police agencies quickly develop for alarm calls that continually turn up nothing.

Four:
A phone conversation

Call the subscriber. Either the subscriber or a representative should meet the police at the premises and let them in for searching, arresting, etc., and relock his building. (Because a service representative of the alarm company is usually required to restore any damage to the alarm system and to reset the alarm for the balance of the evening, some subscribers contract to leave keys with the alarm company so they don't have to be there. The alarm company personnel represent them at the premises, admitting and accompanying police and relocking. There is usually a charge for this added responsibility of caring for keys.)

Five:
Follow-up

Finally, dispatch an alarm company service vehicle. Alarm company employees will, if authorized, represent the subscriber when necessary, reset the alarm system and, if possible, repair any damage that may have occurred in the system. Again, the record of prior service calls and possible equipment malfunctions provided instantly by computer can save hours or days in locating and correcting a fault.

Special instructions from individual subscribers may alter or vary these steps in a number of ways, but this is the basic course of action. The subscriber can arrange for any form of service he needs by the instructions that he puts on file at the central station. Some ask only to be notified. Others, in the event of a break-in or burglary, may require that armed guards be posted at the premises until dismissed by an executive of the company. Or, instructions may require immediate notifications of certain persons in addition to the owner, such as the insurance representative.

Each detail of the action taken and all reported events are permanently recorded in the central station data bank and, if later needed in court, are valuable evidence of what happened and when.

Your keys...

A Note on Keys: Practices for handling subscriber keys in the alarm industry range from truly negligent to reasonably adequate. UL has determined standards for handling subscriber keys that require, among other things, that keys be kept in coded envelopes in locked boxes in the service vehicles. The master code, identifying which key belongs to which subscriber, is kept in the central station in the hands of a supervisor who gives the code for a specific location to the service man when he calls in from the field. After use, the keys are turned in to be re-coded. In addition to that, each use of subscriber keys is logged; not a fail-safe system, but fairly reliable. It does not prevent a dishonest serviceman from duplicating a key, but it does prevent the theft of a whole box of keys. While alarm service personnel are usually bonded, the occasion of a dishonest service person can occur, so caution is advised.

...a heavy responsibility

Private Patrol Services

An option...

In the relationship between the alarm company and the local police agency, the problem of false alarms has always been of great concern. With the rise of many smaller alarm companies, the problem has reached epidemic proportions. In an attempt to force alarm companies to control the problem, many cities have adopted policies of fining alarm companies and/or their subscribers for excessive false alarms.

As noted above, this creates a situation in which decisions sometimes have to be made by the central station supervisor when alarms are received from problem alarm systems. For this reason, the full-line central station alarm company maintains its own armed patrol force licensed to act as a private police agency. A patrol vehicle can be dispatched ahead of the police to investigate whenever there is strong reason to believe an alarm is a false one.

...to prevent false alarms

Sometimes, an outside patrol service is contracted to perform this duty. Whenever this is the case, it should be considered essential for that service to have all patrol vehicles equipped with two-way radio communication to a central dispatcher. Some private patrol services have been known to maintain communication by pay telephone, a method which cannot be counted on.

A word of caution: Only a licensed patrolman is qualified to carry a weapon and perform the duty of a private law enforcement officer. On the other hand, alarm company service personnel, dispatched after an alarm is received, are there primarily to restore or repair the alarm system.

Rending The Veil of Illusion

As we leave the subject of the central station, we would like to dispel a pervasive illusion that unfortunately surrounds the alarm industry.

Because the alarm industry combines police work and high technology, it tends to acquire an aura of glamour based on the perception of the central station as something like a CIA spy headquarters, loaded with ultrasophisticated technology and able to perform functions that are, in fact, totally beyond its capacity.

**What the alarm
industry is not...**

While the modern, well organized and well run central station is a fascinating nerve center, handling complex and diversified responsibilities, it is, like yours, just another business with an enormous amount of daily pressure. Its effectiveness depends upon how well that pressure is being handled.

If you are deciding about alarms and have never seen a central station, a visit to one will prove illuminating and instructive. We highly recommend that you do this before choosing an alarm company.

Diagram of a Basic Alarm System

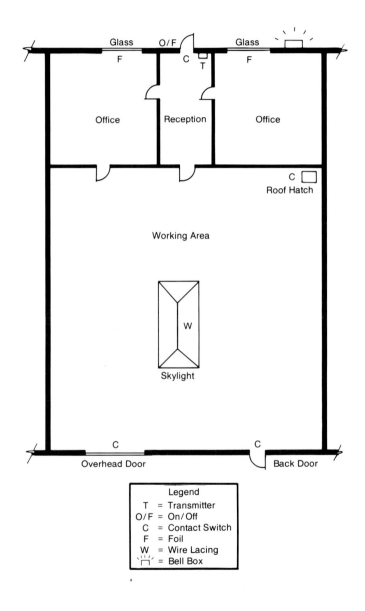

BASIC BURGLAR ALARM PROTECTION

We covered the central station first because it is the heart and mind of electronic protection, and a clear picture of what is going on there is essential to understanding how security is provided.

The basic burglar alarm package...
When you design an alarm for your business, the central station of the company you choose is already in place, providing its services to thousands of subscribers over a wide area. But, in your own building, you will be starting from scratch to plan an electronic installation and operating procedure specifically suited to your needs. Your decisions define the security you will receive and the costs you will pay.

...does not exist
No alarm service company has yet found a way to package a neat 'Basic Burglar Alarm System' that works equally well for a large number of subscribers. Different kinds of property to be protected and different types of buildings, as well as other variables, make every alarm installation different. You and a representative of your alarm company will have to sit down together to develop your specific alarm system, deciding what alarm equipment will work best for you. In every case, the system will begin with a number of basic elements needed in every security system. Working from the general to the

specific, you and he will be defining your alarm system as you go, probably by asking one important question...

What's The Very Least I Need?

But for every situation there is...

This is the uppermost question for about 75% of all homeowners and business owners the first time they apply for a burglar alarm system.

Most ask the question bluntly at the outset and make it firmly known that nothing else is of any concern. Others ask a few minor questions without hearing the answers before getting around to, 'Well, what's the least it takes for my place?' Of course, there are always those who want the best that can be had, regardless of cost.

...a minimum security system.

But let's start our description of a basic alarm system by answering the familiar question, 'What's the least?' That will be our starting point.

The Minimum Alarm: An Example

Building a minimum alarm system is a process of balancing practical benefits against cost.

Perimeter protection:

The minimum alarm system usually involves what is called perimeter protection...meaning the installation of an alarm trip-point at every exterior opening that could be used by an intruder to gain entry into a building. Each trip-point must be designed so that it will initiate an alarm signal if the opening it is protecting is broken, jimmied or unlocked at an unauthorized time. This approach has the advantage of providing the earliest possible initiation of the critical response cycle.

To describe one 'very least' burglar alarm system as an example, we'll hypothesize a small business of relatively low risk occupying a building that has:

The points of entry • • • • •   • one front single pedestrian door, solid wood

• one rear single pedestrian door, metal

• one rear overhead door, metal

• 12' plate glass on each side of front door

• one roof hatch, metal

• one skylight, 4' x 6'

We have chosen the example of an ordinary business in a downtown area not because it is typical, but because it is quite simple, representing the ideal situation to demonstrate the concept of 'minimum' alarm system. Anyone planning to have an alarm system installed in a home will find that the 'minimum' system is considerably more complex—even if there is less area to protect. That is because, typically, a dwelling presents more openings to the outside and a more complex pattern of living. We will examine these differences in detail in a chapter called 'Home Alarms.' But, even those readers who are primarily concerned with protection of a home should study this example carefully. No matter what kind of building is being protected, the same fundamental methods apply. Alarm companies were protecting businesses like these long before there was a market for home alarms. The methods they use in homes today evolved from their experience in businesses like these. And the diligent application of these methods represents the best security that can be provided in any situation.

In our example, the primary consideration is that every one of the items listed above is a possible point of entry for burglary. Each of these points must be protected by some kind of alarm trip-point.

Now let us assume that the owner of this business applies for insurance and the insurance company informs him that he must have an adequate burglar alarm system to be eligible for the policy he desires.

The owner then informs his alarm company that he wants the 'least possible' burglar alarm system. What will it be? When cutting protection to the minimum, one must juggle sacrificed security with increased risk. How far can we go?

One thing is obvious: each of the building's entry points listed above must be an alarm trip-point. Leaving any of these points unprotected is leaving an open door for burglary. This minimum necessity is readily agreed to by everyone concerned.

It is conceivable that a practical minimum system could do nothing more than ring a local bell on the premises. While the ringing alarm bell will deter some burglars, it will not usually bring a response because most people ignore it. For this reason, our model of a minimum security system must ring a bell and send an alarm signal on to a central receiving station. This arrangement is generally required by insurance companies before they will insure for burglary. The addition of a basic central station connection does not add significantly to the cost.

So let us begin by designing for this business a basic system that covers just these entry points. How can this be done in the least expensive way?

The lowest cost installation for our example would be something like this:

The detection devices • • • • •

- one magnetic contact switch (single pedestrian door, front)

- one magnetic contact switch (single pedestrian door, rear)

- one floor contact switch (rear overhead door)

- metal foiling (two front plate glass windows)

- one magnetic contact switch (roof hatch)

- wire lacing (skylight)

- wire and connectors as needed for installation

The support elements ••••• These trip-points must all be wired to...

- one On/Off switch (key operated)

- one local bell (on outside front wall)

- one silent transmitter

- one telephone line connection

In our example, the silent transmitter is an automatic dialing device (digital dialer) which uses your existing telephone line without adding additional monthly telephone costs.

The minimum cost ••••• To install and service the sample system described here would cost an estimated:

Initial installation .... $400 to $600

Monthly service charge .... $35 to $55

The installation price is for the labor of installing the equipment and incidental materials.

The monthly service charge includes amortization of equipment, all telephone line charges (none in this case as your existing phone line is being used), maintenance and service of equipment, responses to alarms, monitoring and logging of all activity. Alarm equipment usually remains the property of the alarm company (though it may be purchased outright by a subscriber in some instances).

Anyone fairly appreciative of comparative costs of materials and labor in today's economy may suspect that these charges are smaller than those for other types of installations and services. They are—and this is possible because alarm companies typically use a three-to-five-year period as the standard term of the service agreement, thus making it possible to achieve long-range amortization of installation and equipment costs and profit which would otherwise be prohibitive for many small businesses.

A contractual obligation

A note on contracts: Be aware that your service agreement is a contract which obligates you to pay for your alarm equipment over the term of the agreement. In effect, you are entering into a three-to-five-year payment plan which does not expire if you go out of business, move to a new building, or decide you no longer want the alarm system before the term is over. Because alarm companies tend to operate on tight cash flows, many smaller companies sell their individual contracts to financial institutions to get money for equipment purchases. You may find that you are indebted to the Bank of America rather than your local alarm company. See our section on contracts for further discussion of alarm agreements.

Is A Minimum Alarm System Acceptable?

In the system described in the example, the minimum number of trip-points of the simplest type are installed only at the obvious possible entrances. The system is then connected with the central station by telephone line. In the past, this was normally a special circuit called the McCulloh Loop connection, but the ordinary telephone line is fast becoming a common form of line connection.

Does it satisfy you?

Will this system give adequate protection? Will it be acceptable to the insurance company? The answer to both questions is 'perhaps yes'...not a strong and definite yes, but a fairly good probability. As a matter of fact, a great number of burglar alarm systems now in use (in the same general category as our example) do have such a minimum system and are accepted by insurance companies.

Does it satisfy your insurance carrier?

The question, remember, is 'What is the least I can get away with?' And by this date the alarm companies, the insurance companies, and the subscribers have been at it for many years, working out the 'minimum' that would be barely acceptable to everyone, depending upon the value of what is to be protected.

We must warn you not to suppose that such a minimum burglar alarm system either would or should be acceptable to all insurance companies.

Does it meet the need?

Much depends upon whether the alarm company central station is UL Approved or whether the company follows a policy of building every subscriber's alarm system to UL Specifications even when the insurance company does not require UL Certification of that subscriber's system.

In general, when considering any particular burglar alarm system for approval, the insurance underwriter tends to give weight to the reputation of the alarm company. If the alarm company has earned the respect and approval of the insurance company, the underwriter is often satisfied to leave the details of the installation and the choice of equipment to the alarm company.

Procedures For Minimum Security

Good procedures make alarms work right.

So far we have described security only in terms of equipment and procedures of the alarm company. While we will continue to do so throughout the book, please remember that your security also depends on the procedures you follow to make the alarm system work as it is intended. This point is important and we will discuss it thoroughly in a section called 'Living With Alarms' later in the book. For now, keep in mind that subscriber procedures go hand-in-hand with the development of good security. Your procedural responsibilities will grow in proportion to the complexity of the alarm system.

Upgrading Security

As you look at options...

For many businesses, the minimum alarm system provides the most thorough protection that can be achieved at a reasonable expense. Other businesses, however, because of greater vulnerability or the presence of merchandise of greater value, cannot be adequately protected so simply.

So you can better understand the possibilities of a modern burglar alarm service, we'll now survey the things that can be done to strengthen and upgrade an alarm system.

...keep a sense of perspective...

In more advanced alarm installations, the added cost must always be balanced against the potential gain in security. Often, however, improving an alarm system does not mean substantially increasing its cost.

There is more to protection than alarms...

In many cases, the installation of more sophisticated, and therefore more expensive, devices may allow an area to be protected with fewer trip-points at an only slightly greater total cost. For example, one ultrasonic unit may eliminate foil on a number of small factory windows, reducing installation and/or monthly service charges.

Anchor Pad

Before we leave the topic of minimum security, we would like to point out a hidden factor that can make an alarm system remarkably more effective without adding substantially to its cost.

The most sophisticated or expensive burglar alarm system cannot prevent a successful burglary if valuable items are left in easy grasp of the burglar. An inexpensive bolt-down kit may be more effective than costly additions to an alarm system in protecting an office in which the most likely target of theft is half a dozen expensive typewriters.

Cablelock

SLOW HIM DOWN!

A simple bolt-down device, manufactured by Cable-lock Office Equipment Security of Santa Monica, Ca., consists of a small cylinder lock secured to a hole drilled in the desk or table top. When locked in place, the cylinder retains a steel cable that is permanently bonded at the other end to typewriter or computer. The equipment can be unlocked at any time and moved to another location. But, when secured, it can only be moved as a unit with the entire table or desk. Varieties of these locks sell for $60 and up.

A more elaborate product that can be installed in minutes without drilling holes in office furniture is marketed by Anchor Pad International. The Anchor Pad is a two-part platform that rests under the office machine. A special glue bonds one side of the pad to the desk surface. The other side bolts to the under-side of the equipment. The pads are interchangeable so that equipment can be moved from one pad to another. However, a special solvent is required to remove a pad once it is bonded. The standard pad for one-piece equipment retails for about $125 to $190, including installation. Units designed for component computer systems sell for $330 to $350.

By itself, such a device is not a guarantee against theft. Even the best one can probably be defeated by a determined thief. But, taken as one part of your overall protection plan, it can substantially reduce your exposure to loss by turning the Cycles of Security more in your favor.

Remember:

Receiving notification that someone is burglarizing your premises is important. Having him still there, trying to complete his purpose when the police arrive is also important. The principle here is, 'SLOW HIM DOWN.'

BALANCING PROTECTION AND COST

The tools of security are as many and complex as the properties the alarm company must protect. Every home floorplan, every type of business, every piece of merchandise to be protected, every family, every special problem requires a slightly different application of security equipment.

As we mentioned already, the design of every alarm system is an exercise in balancing a level of protection with its cost.

This exercise begins with the critical question...

How Elaborate and How Expensive?

The best guideline you can have in making this difficult but fundamental alarm decision is a well-studied knowledge of your level of risk.

There is no ready formula to determine risk. Rather, risk is a relative concept growing out of numerous factors which are different for every place and type of business and are sometimes impossible to measure except by intuition and experience. In the home, risk becomes even harder to define, as it is tied to the perception of personal safety and subjective values vary dramatically from one person to the next.

To design your alarm you need to assess...

How much could you lose?

In every case, the evaluation of risk must be determined between you and the alarm engineer, who, remember, is not usually an engineer at all, but a salesman who has experience in designing alarm systems. Spend some time to define for yourself what you think your risk is and make this clear to the alarm engineer. Secondly, make a serious attempt to set rough parameters for the budget, taking your perception of the risk into account. If you will do this, the development of your alarm system will proceed much more smoothly.

How easily could you lose it?

Your alarm company engineer is going to be looking at your risk and budget from a different perspective. Generally, as he walks onto your property for the first time, his interest will focus first on the type of goods you are trying to protect. Usually he will frame this question in terms of dollar value, rather than human value. While you may find that somewhat disconcerting, remember that whatever threat you face is probably the result of a similar viewpoint.

He will want to know the total value of your inventory or household belongings, as well as the value per item, and the ease with which each unit of value could be removed.

Beyond that, he will be looking into his own experience for comparative information that may not be known to you. He will be thinking about the demand for your inventory or belongings on the stolen goods market and he will also be asking himself what kind of burglar would be attracted by the particular type of goods you have.

What are the indirect losses?

A more subjective, but often very important, question he will want to know is just how much your business could be harmed by a successful robbery or by damage to crucial equipment or records.

Some companies, to be sure, are in a position that a single intrusion would appear intolerable. A com-

And...

pany that prints negotiable banknotes will go to almost any lengths to gain that elusive concept of 'absolute' protection. The nature of its business demands nothing less than the highest level of vigilance.

Similarly, a company which could lose $100,000 of business by the destructive acts of a vandal shutting off power to a computer room might require a more elaborate approach to security than a similar company in which the primary goal is preventing the theft of a $100,000 piece of equipment which would take several hours to cart off.

What would it cost to prevent the loss?

The next question which will occur to the alarm engineer is what type of burglar would be attracted by your business. Will he be a neighborhood thief who snatches and runs or a sophisticated break-in artist who studies your alarm system and plans out his attack?

And, the engineer will make a thorough examination of your building in search of the factors which make up its vulnerability to loss.

A few such factors are:

Location	Local crime rate Level of nighttime activity Existing local security
Exposure	Ease of access to building Nighttime lighting Number of employees with keys
Type of structure	Light office Factory Warehouse Outside yard areas
Weak spots	Weak wall with neighbor Common drop ceiling Number/location of windows

Finally, the engineer will consider several special factors such as the quickness of police response in the area, possible danger to life, the extent of exposure to fire and the likelihood of vandalism or armed attack.

From these questions, the broad guidelines of your commitment to a security system will be established. When the engineer knows the outside limits of what you expect to pay he can then begin looking at the specific items in his repertory which will give the best service within those limits. The final plan will consist of many compromises between the value of protecting specific weak points and the cost of doing so. As you may not agree with every choice he is making, we point this out so that you can enter into discussion over the choices that make up his proposal. In most cases a completely different approach could be taken within the same cost framework.

At this point we would suggest that to help in assessing your risk you seek the professional judgement of your local police and fire agencies, your insurance underwriter, your alarm company and perhaps even your accountant. Clear thinking and common sense are vital. You must understand your actual, not theoretical, risks to achieve adequate protection.

In addition, we think it is a good idea, when planning an alarm system, to get estimates from two or more companies to compare cost and to see if one may come up with a better idea than another. In comparing bids, make a simple floorplan of your building showing all rooms and points of entry and ask the bidding company to indicate on it all alarm equipment and the path of alarm wiring to every sensing device. This may help to draw to your attention possible installation shortcuts which seem to make one alarm system more economical than another, but ultimately will cost more in service failures and difficulty in operation.

Also, one alarm engineer (or alarm salesman which is more likely the case) may see points that another misses or be able to show you different strategies for reaching the same end as you begin to redefine the question asked earlier, 'How elaborate and how expensive?' and begin to focus on the real question in buying alarms...

What is the Right System For the Risk and Budget?

This analysis generally begins with something like what we have called the basic alarm system and then pursues one or more of the following basic approaches until you feel you have reached an acceptable balance:

**The basic
approaches • • • • •**

- Selecting better types of sensing devices at trip-points.

- Adding more or special types of trip-points for fuller coverage to protect high risk areas or avert dangers.

- Adding alarm devices for armed robbery.

- Increasing security by upgrading the transmission capabilities and the line connection to the central station.

- Increasing central station supervision of the alarm system by changing the alarm format.

- Using UL Certification Standards according to the degree of security needed and insurance recommendations.

- Using optional devices to perform auxiliary services (closed circuit TV or film cameras to visually record intrusion, passcard key systems, etc.).

- Using patrol and guard services.

- Upgrading and strengthening your fire prevention capability as well as potential loss from
  water damage or other industrial hazards.

In the next chapter, called 'The Tools of Security,'
we will discuss the first eight of these nine approaches in detail, reserving the subject of fire protection for a separate chapter because of its distinct
nature.

We propose the next section of the handbook as a
'suggestion box' that might spark just the idea you
need to solve a special problem or strengthen weak
points that might be or have been troublesome.

Everything mentioned or described in this section
of the handbook is available, of course, for use in
any basic alarm system. Someone might conceivably
start with the type of 'minimum' basic system
depicted in the example above and, with time and
growth, gradually upgrade his alarm system in every
one of the above-mentioned ways.

At this point, anyone who is not concerned with the
application of alarms in business might prefer to
jump to the 'Home Alarms' chapter. We would caution, however, that most of the devices and strategies described in 'Tools of Security' are fundamen-
tally applicable to the home as well. We therefore
recommend your careful reading of the next section, either now or later, to enhance your understanding of how the home alarm system works.

THE TOOLS OF SECURITY

At this point, you should have a pretty good idea where your specific situation demands more than than the basic alarm example we gave earlier.

In this section we will describe the wide variety of equipment and services available to you.

Because the material here is so complex, we have broken it into eight major subheads which correspond to the eight approaches to improved security listed above, with the exception of fire.

To reiterate, they are:

1) **Types of Sensing Devices**
2) **Special Trip-Points, Vaults and Safes**
3) **Holdup Alarm Systems**
4) **Transmitting Devices and Connecting Lines**
5) **Supervision**
6) **UL Certification**
7) **Auxiliary Devices,** and
8) **Patrol and Guard Services**

As you begin to contemplate the dizzying array of alarm equipment available today and the diverse uses that can be made of it, we would like to touch upon one element of good alarm design that will help keep you within the realm of practical management.

Zoning Alarm Circuits to Keep the System Simple

Zone Indicator

As an alarm system grows more complex, so does the trouble it can cause if it is poorly designed. Eventually, a point is reached where the very complexity begins to interfere with its efficiency. At this point, a substantial benefit is achieved by breaking the system into smaller systems called zones, each with its own circuit to the alarm transmitter. Separate lights on the transmitter for each zone will greatly speed the task of finding the door that is not closed or the box that is blocking a light sensor as you leave for the evening. Even more important, careful zoning helps your serviceman track down problems when they occur. In conjunction with more elaborate alarm transmitters, zoning can also be used to tell the central station what type of alarm device is signaling and where on your premises it is, speeding police or fire department response.

And now, the Tools of Security...

1. Types of Sensing Devices

Integrity starts with the cheapest switch...

Much of the success of any burglar alarm system rests upon the quality and appropriateness of the trip-points that actuate the alarm signals. The devices used must be of high quality as they must be reliable over a long period of time and under all conditions of weather and wear. Great losses are suffered due to devices that rust out, wear out, break easily, cause false alarms, or fail to operate. UL Approved devices are recommended in all cases. (Remember, UL Approval of a specific piece of equipment does not include approval of the whole alarm system or the central station.)

The devices must also be appropriately chosen for

the specific duties they are to perform. Some devices are too sensitive for general use and some too highly specialized (one of the most common causes of false alarms). An intricate and expensive device that provides an excellent trip-point in one application may be, and often is, totally unsuited for another.

Quality alarm companies buy quality components and apply them correctly.

Upgrading and strengthening an alarm system is sometimes done merely by using a better sensing device or combination of devices at a critical point. It is, therefore, helpful to have a general awareness of the many types of sensors available.

As alarm devices are added, attention should be given to a subtle but important concept—the ability, or inability, of a device to report if it is not working properly. This feature is incorporated directly into some alarm devices, but is more often a factor of the way the alarm circuit is wired.

The way a device is wired can make a difference.

Sensing devices can be wired two ways: 'normally closed' and 'normally open.' The 'normally closed' type is monitored for a continuous flow of electrical current through the circuit. When the device is triggered a switch opens, breaking the flow of current (opening the circuit). The 'normally open' type does the opposite. An alarm is generated when a switch closes in the circuit, starting a flow of current. Most alarm systems will incorporate both methods with occasionally a further variation measuring the change of resistance in the circuit.

However, the 'normally closed' configuration is inherently more reliable because a failure will usually show up as an interruption of current flow, a condition which demands attention. A 'normally open' switch, on the other hand, could stick open or break without showing a change in the circuit status. As technical as this may sound, it will help for you to know how each trip-point in your system works to facilitate tracking problems when they occur.

The best type of alarm circuit is a supervised circuit that incorporates an end-of-line resistor at the point farthest from the alarm transmitter. This establishes a specific electrical resistance in the circuit that is monitored electronically in the transmitter at all times. Any variation in this resistance is interpreted as an alarm condition.

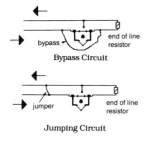

Bypass Circuit

Jumping Circuit

Quality is not always visible...

The most secure method of connecting an alarm device into the alarm circuit is to use what is called a 'form C' contact switch. The 'form C' switch is configured to 'open' the circuit containing the end-of-line resistor and in the same action 'close' a new circuit back to the transmitter without the end-of-line resistor. This prevents bypassing the alarm device containing the switch or shorting the circuit with a jumper. A bypass around the device is negated the moment the alarm device is tripped, closing the new circuit back to the transmitter. Jumping the circuit across cuts out the end-of-line resistor, causing an immediate alarm.

The simplest switches open or close an electrical circuit mechanically. More sophisticated devices use small relays or solid-state electronics to open or close a circuit. Sticking relays or failure of a poorly made microchip are common causes of false alarms or alarm failure. Thus you can see that skimping on one of the least expensive parts, or the method of wiring it, can compromise the whole system.

Active vs. Passive

A related consideration is the distinction between devices which are 'active' and those which are 'passive' in their method of detecting an alarm condition. 'Active' detectors, such as ultrasonic, radar, microwave and capacitance units, create a standing field of energy and measure the change of activity in that energy. A 'passive' detector, such as infrared, sound or phonic detector, vibration detector or heat detector, monitors the normal condition of the environment within its range and alarms when something in the environment changes.

Quality alarm companies always take these problems into account during installation, sometimes necessitating greater effort and expense.

Now, without being too technical, we will list the various types of sensing devices with a brief explanation of each.

One way to understand burglar alarm sensing devices is to categorize them by the way they are tripped. Trip switches may be actuated by mechanical contact, circuit interruption, interrupted light beam, reflected energy, infrared radiation, capacitance disturbance, vibration, sound, pressure, or heat. We will take these devices in that order.

Mechanical Contact

All mechanical contact switches operate by opening, closing, or changing the balance of an electrical circuit in order to actuate an alarm signal. The majority of doors, windows, and ordinary openings are best served by simple contact switches. The variety of contact switches includes:

Basic to all security systems ...

Magnetic Contact

Magnetic Contact. For pedestrian doors, windows, hatches, sliding doors, etc. A floor-type protects overhead, garage, or roll-up doors. This device is tripped not by actual contact, but by the pull of a magnetic field on a sealed reed switch, causing it to open or close an electrical circuit. These come in a variety of configurations including some that can be recessed within the door jamb and are completely hidden from sight.

High-Security Magnetic Contact. A more complex version of the magnetic contact with anti-tamper protective features.

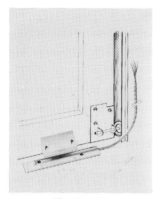

Floor Contact

Mercury-Flow Switch. A device that works like a level and is used on doors and windows that tilt open. Mercury resting between two electrical contacts is disturbed by movement, opening or closing a circuit.

Button Switch. A mechanical device sometimes used on hinged jambs of doors and windows. This device is generally considered obsolete.

Tamper Switch. A button or magnetic switch installed inside an equipment housing that is actuated by an attempt to tamper with the housing.

Safe Switch. A high-security tamper-proof switch made to UL Specifications especially for safes.

Pressure Switch. Used in mats, under rugs, etc., that can be actuated either by the weight of a footstep or the loss of pressure when anything over it is removed.

Button Switch

Ball-Trap, or Clip-Trap Switch. A switch that is actuated by the removal of a metal part from between two springs. This device is also generally considered obsolete.

Damage-Induced Circuit Interruption

Sometimes current is interrupted by the breakage of a wire, a foil strip, a screen, etc. Devices of this type, unlike those discussed under Mechanical Contact above, do not reset themselves after once being tripped. A service call is mandatory after any alarm generated by the following devices:

A practical standby . . .

. . . aesthetics aside

Foil. Three-eighths-inch wide strips of metal foil are applied near the edges of window or door glass and a current is passed through the foil. This foil is very easily broken and separates quickly causing a signal if the glass is cracked or broken.

High visibility ...

The great value of foil, and one of the main reasons for its extensive use in alarm installations is that it places alarm protection at the perimeter of your business and is low in material cost.

While foil is practical and often simple to install, there can be a lot of hidden labor hours in its installation, and it is not as trouble free as it would appear.

...can be a deterrent.

Openable windows, sliding glass doors, and doors with glass in them can also be foiled. With these, a flexible device is used to carry the alarm circuit to the moveable portion of the door or window. In multiple-pane windows, a bridging strap has to be installed to carry the circuit from one pane to the next.

Foil is placed on glass and varnished over to protect it against damage and hold it to the glass. Because most windows are exposed to the sun, the foil and varnish heat up and expand during the day and cool down and contract at night, creating a tension on the foil and all of the connecting points between the foil and the electrical circuit. This will eventually develop into a service problem.

In addition, the wear and tear usually caused by window washers, window display dressers, and the public (especially on glass entry and exit doors) leaves damaged spots which must be patched with foil strips, causing the addition of two new connections which are inherently weaker than normal foil.

UL Specifications require, among other things, that foil be installed on three sides of every protected pane, unless very small. However, to cut the cost of installing foil, some alarm companies will use a process called 'strip-foiling,' which means running a single strip of foil across a window pane. This sometimes results in one company's bid being lower than another's. This is not necessarily a bad technique on a small enough pane of glass. But foil installed this way has to be redone if, in the future, you decide to have the system UL Certified. 'Strip foil' is allowed by UL on tempered glass, as any break in the glass will cause the entire pane to shatter.

Only where really needed

Wire Lacing. Tightly stretched, tempered wire carrying a slight current is laced across openings (vents, skylights, etc.) or over paneling, ceilings, thin walls, or other places vulnerable to penetration by an intruder. These wires are laced so that they are easily broken if displaced by penetration of the wall, causing an alarm to be set off. In UL Approved systems, two electrical circuits are interlaced so there is less chance a burglar can defeat the circuit by jumping the wire lacing before cutting the screen, undetected.

Screens. Two types of special alarm screens are available as protective devices for windows and similar areas: **wood dowel** screens and **aluminum** screens.

Wood dowel screens are not actually screens, but are wood lattices with six-inch square interstices. The horizontals are square wood strips and the verticals are hollow dowels containing fine wires through their centers. These are wired into the alarm circuit. Though seldom installed any longer, wood dowel screens remain in widespread use and are very effective if properly installed.

Wood Dowel Screen

Aluminum Screen

The aluminum screen is, to all appearances, an ordinary aluminum frame screen with nylon mesh such as may be seen on any business or residence window. However, current-carrying wires are woven into the screen mesh at four-inch intervals and the frame inserts into contact points at both top and bottom so that either removal of the frame or cutting of the screen interrupts the alarm circuit current and actuates a signal.

The dowel screens are highly effective and are approved for use in UL Certified systems. The aluminum screens are most often used where appearance is a consideration. They are custom made to fit any window and do not detract from the appearance of the building. Some double-laced aluminum screens are UL Approved for burglary protection.

Panels. Wire lacing concealed between two pieces of paneling (such as composition board) can be made to fit to doors or walls of any material other than glass. Any effort to penetrate the door or wall breaks the wire, setting off an alarm.

A new, and highly reliable, version of the panel uses a lamination of two plastic sheets, each approximately three-eights of an inch thick, with wire laced into channels approximately eight inches apart. Sheathing of the wire inside the plastic keeps damage to a minimum. The panel can be left transparent or painted.

A note on lacing: Instead of using panels, some companies will lace a wire back and forth across a wall or ceiling, wrapping each end around a nail. Such lacing was commonly installed 30 to 50 years ago, but is today considered obsolete by most companies because of the exposure to damage and the susceptibility to false alarms caused by shorting to ground from moisture.

Interruption of Light Beam

This principle is simple: a thin beam of invisible light is projected onto a receiving unit with a photoelectric cell that maintains an electrical circuit as long as the light falls upon it but actuates a signal if the light is interrupted. Two devices of this type are the **High-Intensity Light** (often referred to erroneously as a 'Laser Beam,' but more accurately a light-

The most reliable space protection

emitting, gallium arsenide diode) and the **Photo-electric Beam.**

High-Intensity Light. A light emitting diode (not a laser beam) makes a trip-switch of a beam up to 1,000 feet long—a line that cannot be crossed by an intruder without causing a signal. The beam is invisible, sometimes pulsed, and cannot be defeated easily. The units are small and sturdy and have their own standby power supply so that short power failures do not disrupt their effectiveness or cause false alarms.

'Laser'

Photoelectric Beam. Usually uses an incandescent light (filtered to the invisible infrared range) which sends a beam 60 to 500 feet. Momentary power lapses (when standby power is lacking) and burnt out bulbs cause false alarms. This device is not generally used in new systems since the advent of beam devices using the light-emitting diode, which do not have the limitations of the photoelectric unit. However, photoelectric units may still be used, despite their limitations. They cost less, are sometimes suited to special applications, and can be designed to be reliable.

Sometimes light beams are configured to create a 'fence' barrier with two to four beams criss-crossed through a rectangular area five to six feet high. In outdoor areas, the light-fence can be set so that two or more beams must be broken to actuate an alarm. This helps to prevent windblown paper and small animals from setting off alarms.

Reflected Energy

The exotics

Various manufacturers provide devices which use reflected energy to actuate a signal. They generally fall into three categories: 1) **Ultrasonic Detectors;** 2) **Radar Detectors,** and 3) **Microwave Detectors.** Although these three types of sensors employ the same basic principle of detecting changes in echoed waves (commonly called Doppler Shift principle upon which sonar in submarines is based), they differ widely in sensitivity and application. Again, UL Approval of any particular piece of equipment is a valuable guide.

The common virtue of all these sensors is in their ability to make a trip-point of a volume of space. In some situations this is infinitely more desirable than a trip-point consisting of a single point (contact switch) or a long line (light beam).

A note on False Alarms: The most common source of false alarms with this equipment is having the sensitivity level, or volume, too high for the application.

For very stable environments

Ultrasonic Detectors. The ultrasonic unit consists of a transmitter and a receiver which may be in the same housing or in two separate housings located a distance apart. The volume protected varies considerably according to the type of device and the surfaces presented in the room—its walls, furnishings, etc. Soft surfaces, carpets, drapes, etc., absorb and diminish ultrasonic sound, while hard surfaces reflect it over greater distances. The volume of protection also varies with the quality of manufacture of the specific unit, and with its placement in relation to other units in multiple installations. An average unit covers the volume of a 25' by 25' room, but the volume that may be covered varies greatly.

Ultrasonic Transceiver

The transmitter emits a high-frequency tone above the audible level, in the 25-45 kilocycle range. An alarm signal is actuated if the receiver notes any change in this tone. A change would be caused by the movement of an object deflecting the sound waves. Depending on the sensitivity of the ultrasonic device used, a change in tone can be detected from as slight a movement as the disturbance of air caused by an intruder.

There are many models...

Ultrasonic Transducers

...many applications...

Ultrasonic Transceiver

...and many false alarms.

In the use of ultrasonic motion detectors it has always been a difficult problem to create a system that is sufficiently sensitive to be effective yet stable enough to resist false alarms due to ambient noise and vibration.

Ultrasonic motion detectors are excellent and highly effective devices, but can be quite unsatisfactory because of a high false alarm rate when the equipment used is of inferior quality or when the devices are misapplied, or improperly installed. Ultrasonic devices are often misapplied when used in areas subject to air turbulence, noise or vibration (such as from nearby train or truck routes or jet flight paths). Such unsuspected sources as humidity can also cause false alarms. Every installation requires careful engineering to locate the sonic units at properly balanced distances and on suitable surfaces. Ultrasonic devices are made by many manufacturers, some of which, unfortunately, offer an unsatisfactory product.

Nevertheless, when an ultrasonic detector is well made, correctly installed, and used in a proper environment, it contributes to an especially effective alarm system.

The development of microchip circuitry and sophisticated electronic logic systems has produced a superior variation of the ultrasonic detector able to filter out many of the disturbances that an ordinary unit would have perceived as an intruder, taking note only of changes in the mean distances between objects in a room. Anything moving toward or away from the device in the space volume it covers would actuate an alarm.

This device, called **Range-Change Ultrasonic Detector**, has proved less susceptible to the high rate of false alarms experienced with unfiltered types. But take note: all such devices are by nature extremely

Be careful.

sensitive and must not only be carefully installed, but also must be watched over by good security engineers during early weeks of use to be sure that any hidden problems are found and solved.

A note on noise: Some early models of ultrasonic devices used tones at frequencies as low as 18,000 cycles per second, inaudible to most people, but within the hearing range of some. Those who do hear, or almost hear, this sound usually find it irritating. Many employees refuse to work in its presence and customers sometimes refuse to remain in stores when they hear this extremely high-pitched tone. The better quality models now available use tones above 25,000 cps which do not give rise to this problem.

Radar and Microwave require a high level of engineering capability...

Radar Detectors. Sensing devices based on the use of radar waves are similar in operation to ultrasonic detectors but, since they use a much higher frequency wave, can cover greater space volumes and perform with extra capabilities.

Their extra sensitivities might eventually offer a way to protect outside areas in addition to large interiors. However, the radar detectors now available have failed to meet the requirements of many alarm companies due to an excessive false alarm rate. This is very likely due to the fact that they require extremely careful installation engineering. Caution is advised.

...and a high level of alarm company support.

Microwave Detectors. The highest frequencies of all are used by microwave detectors, which operate at millions of cycles per second. Although the operating principle is essentially similar to the ultrasonics, the range and sensitivity is considerably greater.

Microwave fields can be shaped to sense changes that occur within the volume of a large factory or auditorium and can be channeled as an aisle-wide 'beam' in a long warehouse.

Because of the extra keen sensitivity of microwave detectors their use is limited somewhat and requires meticulous engineering. Microwaves are so strong and sensitive that their range may not always end within the desired limits. They may, if not properly installed, pick up buses and trucks passing outside or other events causing unintended strong vibrations. This characteristic can sometimes be used to advantage by setting up a detection field through a wall or several partitions. In this case the antenna controlling the range and intensity of the microwave radiation must be carefully balanced or trouble will arise. Microwave detectors usually respond when a preset percentage of the energy going out is reflected back. Thus, if set so a man at sixty feet would reflect the needed amount, a bus outside the building might reflect the same amount at 200 feet.

Microwave Detector

Microwaves excel in large warehouse applications. However, they are virtually impossible to use in metal buildings because they can interpret the slight warping of a cooling metal wall at night as an alarm condition. Also, metal reflects microwaves and creates a supercharged energy level in the building interior.

In general, all sensing devices using the principle of reflected energy waves require more-than-usual care in application. Qualified engineering and installation of any alarm system employing ultrasonic, radar or microwave detectors is vital to the system's success.

In warehouses, be aware, changes in inventory affect these devices. Be sure you discuss changes in the contents of protected rooms with your alarm company.

Most reflected energy devices come as self-contained

units or with multiple sensors connected to a master control. Usually, the sensors are in pairs with a sender and receiver located some distance off. Up to 100 sensors can be connected to one master control. However, zoning is recommended so defective units can be located.

Not every alarm company knows how to use this equipment correctly. With large and complicated alarm systems, it is to your advantage to choose an alarm company that has a qualified engineering department, even if the costs are higher.

Although alarm system coverage of exterior areas and storage yards is generally impractical due to the many outdoor causes of false alarms, in situations where outdoor protection is possible, the microwave detector is an excellent device. Many times careful engineering and experienced installers can surmount the difficult problems of outdoor alarm coverage and, if the need is sufficiently urgent, it may be attempted.

A note on microwave: A recent development in microwave, called Point-to-Point Microwave, allows vastly improved outdoor coverage at a considerable increase in cost. It will be covered thoroughly later under Special Trip-Points/Outside Areas.

One of the great values of microwave and radar is that because of their high frequency, they simply do not see the usual things that bother ultrasonic type devices. Wind, rain, fog, dust, sound, all fall below their range of visibility.

Infrared Radiation

A good alternative to reflected energy

When it is necessary to use a sensor that is not affected by either sound vibration or air movement, as the ultrasonic type sensors are, the infrared detectors are highly effective. These units, like the motion detectors, provide volumetric coverage. By using multiple detector heads in the system a great variety of locations can be supervised.

Infrared sensors perceive changes in infrared radiation (waves beyond the red end of the visible spectrum) caused by emissions from an intruder entering the protected area. These devices can operate quite reliably, but only if installed to avoid direct exposure to heat from sunlight (through windows, skylights or metal walls), heat-emitting appliances (ovens, water heaters, coffee makers) and floor areas which might be traversed by small animals.

Infrared Detector

Possible sources of trouble for infrared sensors are: incandescent lights that turn on or off automatically (or can burn out while the protection is on), pilot lights, walls or ceilings that heat or cool rapidly with outside temperature change, hot water pipes, etc.

Capacitance

For high values

Capacitance detectors are often used in protection of safes, file cabinets, metal storage cabinets or similar self-contained metal objects. Several such objects can be grouped together on the same circuit, if desired.

Capacitance Alarm

The capacitance detector (also called Proximity Alarm) creates a protective electrostatic field around an insulated metal object. The field is so sensitive that the touching of any part of the protected object causes a signal. The device is made to UL specifications and is approved by UL for protection of safes.

A note on capacitance alarms: The safe to be protected must not touch walls or have direct contact to the floor. Insulation pads must be placed under the safe, a job which often requires a safe moving company.

Vibration

For glass when foil is unacceptable

A vibration detector is a contact transducer which makes a sensing device of any surface to which it is attached.

Electronic Vibration Sensor

This type of sensor is not suitable for most business situations, but excels in highly specialized applications. Vault walls, for example, can be made to actuate a signal if tapped by a hammer or tool. According to the need, vibration detectors can be made so sensitive that they will detect the gentle picking of a lock or silent drilling on a safe or vault. Because of the extreme sensitivity possible, vibration detectors can be useful in the design of unusual systems for protecting specific items of great value.

In high-security applications an effective UL Approved device is the **Window Shock Sensor,** an electronic monitor that responds to the sound of breaking glass through a small sensing device placed on each pane of glass to be protected. These are highly sensitive to the frequency of breaking glass but will not false alarm if the window is bumped or jarred. Window shock sensors are less visible than foil, alarm screens or other devices and are therefore often used in show windows or showcases in locations such as jewelry stores or museums. It is advisable to have these devices UL approved, as there are many window shock sensors on the market that are unreliable.

Mechanical Vibration Sensor

A third type of vibration detector is a small mechanical device sometimes installed at intervals of several feet along a wall or ceiling that could be penetrated. Two tiny contact points carrying the alarm circuit are adjusted to separate, interrupting the flow of current, if the wall vibrates. This device can be troublesome because of its tendency to false alarm due to ambient vibration when set at a level of sensitivity sufficient to detect someone cutting through a wall. Worse, they are victim to one of the major failings of any device that sits passively with closed contacts for months without being tested or operated. The contact points can oxidize together so they will not separate when they should.

Sound

These devices are triggered by sounds picked up by acoustic microphones installed in the area to be protected. The noise of a city makes this type of sensing unworkable for ordinary business use. However, it may be used inside vaults or heavily insulated storage chambers.

A second type of sound device now in use actually relays any sounds picked up back to a speaker at the central station. The operator there can listen in for the sounds of intrusion, then reset the microphone if he hears nothing. A new noise will immediately set it off again. A limitation of this device is that it requires the central station operator to make a decision based only on what he hears. Because UL will not allow discretion in responding to alarms, such audio equipment does not qualify for UL Approval but can be added to a UL Approved system as a supplemental device.

Heat

Heat detectors are tripped when they reach a preset temperature or rise in temperature faster than a preset rate. They are normally used to detect the outbreak of fire. However, they may also be used in burglar alarm systems when installed inside vault and safe doors to detect torch cutting.

Heat Detector

And that concludes our discussion of sensing devices. There are many other kinds of sensing devices too specialized for listing here, and other recently introduced devices as yet unproved, but applicable in some situations. In particular, microchip circuitry has produced great differences in the type and quality of equipment used in the alarm industry. No firm definitions are possible to include all such variations. Discrimination is always required in matching the equipment to the job to be done.

2. Special Trip-Points, Vaults and Safes

An obviously simple way to strengthen or upgrade any alarm system is to add more trip-points, or to add trip-points to protect special items such as safes, especially vulnerable points in a building, or high-risk areas.

Since security, except for a very few good modern architects, has not been an influencing factor in structural design, many businesses have insecure conditions in their buildings. These hard-to-protect spots may not be noticed by the owner of the business at first, but burglars notice, and some unfortunate loss soon may bring the vulnerable sites to the owner's attention.

Vulnerable areas or high-risk problems may be solved simply sometimes by adding special trip-points. Some that should receive special attention are:

Doors

Effective and inexpensive protection

A simple, inexpensive solution to strengthening protection is to add protection to a few inside doors to your alarm system. If someone penetrates your building without setting off the alarm or hides inside until you are gone, an alarm will activate as he moves around and opens these protected doors. (Remember, these doors must be closed and latched every night before the alarm can be set).

Many doors are not adequately protected by contact switches alone, but require an additional sensor to detect entrance in other ways than by opening the door, such as by breaking a glass or thin panel within the door framework. Glass doors may be foiled, like windows, and thin panels may be laced over with high-tensile wire (which breaks easily if disturbed) and a second thin panel added to cover the wire. Plastic panels are especially useful in this situation.

Metal doors, overhead doors, or doors that cannot be adequately protected by a directly attached device, may be protected by having a 'laser beam' or a motion detector unit cover the area just inside the door so that the movement of an intruder is sensed as soon as the door is penetrated.

It is always advisable to add trip-points specifically designed to detect disturbance of cash drawers, jewelry cases, narcotic cabinets, or any such sites of special attraction. Numerous devices are available for this use, which requires special consideration by the designer of the alarm system.

Three common, but sometimes unexpected, avenues of burglary are:

Windows

Some factory windows have numerous small panes, making foiling not economical and impractical to service, and some very long skylights in factory roofs present the same problem. This may be solved by adding 'laser beams' or motion detectors to pick up any entry not tripped by contacts. If the window's size or location makes a screen appropriate, wood dowel type screens or alarm screens may be used.

Windows...

Vulnerable Walls

Some walls are especially vulnerable for their thinness, or because they are no more than partitions made of thin metal or plasterboard. Protection of these is generally similar to that of factory windows, using 'laser beams' or infrared detectors across an area near the side of the wall to be protected. Motion detectors may not be suitable if the walls are metal,

Walls...

if there is ambient air movement near the wall or the stock being protected is close to the wall. If the partition area is not so large that it makes their cost prohibitive, wood dowel screens or alarm screens may be added. Many times plastic panels will blend in with the appearance of the wall.

Vulnerable Ceilings

And ceilings.

Ceilings may also be easy points of entry for burglary, especially some of the popular types of drop ceilings commonly used in remodeling and in fast, inexpensive construction. The vulnerability of many ceilings has given rise to an increased incidence of burglary through this route. Usually, the best protection is to add motion detection devices to cover a layer between the roof and the ceiling material or a layer between the ceiling and the room area below. Again, this is an excellent application for infrared detection.

Safes and Vaults

A safe is simply a steel cabinet made to be as heavy and immovable as is practical for its location. Protection for a safe is provided by a door sensor and an additional device to detect activity around the safe. Unlike vaults, safes are not generally given interior protection.

Usually UL Certified

Special high-security contacts are used on safe doors, and it is advisable to add an electronic vibration detector or a capacitance detector which signals if the safe is touched. Also, an additional separate alarm transmitting device is often used just for the safe, which extends the protection of the safe to regular day hours as well as night and increases the supervision of its entries.

A vault is a permanent structure of iron, steel, concrete, stone, or other strong masonry and its walls, floor and ceiling are part of the building containing it. The door and frame are usually heavy steel with a combination lock.

The floor safe, technically a vault, can also be contacted with a magnetic switch between the lift-off floorplate and the locking safe door.

Because large amounts of money and other valuables are stored in vaults they need extra protection. Not only must a vault door have an alarm trip device, but also the vault interior should contain additional approved sensing devices.

Sound Vault Alarm

Vault doors also use special high-security contact switches plus other types of sensors on the interior walls such as vibration or heat detectors (to detect torch cutting). Sound detectors may be used in the interior.

Safes and vaults can be connected to the general building alarm or can have their own independent supervised transmitters. Each safe and vault instal-lation, however, always requires special attention.

A note on construction: Underwriters Laboratories has a specific set of standards for safes and vaults, rating them for strength of construction and resist-ance to fire. This is not to be confused with the certification of vaults and safes as part of an alarm system, as will be discussed in the section called 'UL Certification.'

Multiple Structures

Separate transmitters work best.

Some businesses have separate buildings where factory or warehouse work is done at hours different from the office hours. One or more additional alarm signal transmitters may be used to allow separate alarm systems for each building or area, making it possible to have the alarm set for one area while off in another. Aside from the regular protection against intrusion from the outside, this helps to restrict factory workers from entering office or storage areas without authorization and can be a major tool in controlling 'inside' thefts.

Hide-Ins, Etc.

Check restrooms when closing.

A 'hide-in' is an employee or other who remains hidden inside the building after closing time to commit burglary after everyone has left. The perimeter system which only covers doors and windows is not effective in this situation, since the burglar does not need to open a door until the last minute when he is ready to leave. By then he has taken what he came for. Although he sets off the alarm when leaving, he can escape quickly before response to the alarm can arrive.

Motion detectors, infrared detectors, laser beams, other inside area sensors, as well as contacts on inside doors, can be tripped without his awareness and he may find police waiting for him as he leaves. If the hide-in is a professional burglar, there is a very good chance he can defeat the alarm system unless it is set up to detect his movements before he can get to it or the alarm circuit is designed to protect against tampering.

Outdoor Areas

Almost never work due to false alarms ...

Outside yard, storage and dock areas have usually been beyond the protection of burglar alarm systems that send the signal off premises and require police response. This is due to the great number and variety of causes of false alarms. To name a few: animals, birds, children, rain, wind, fog, dust, noise, windblown bits of paper, leaves, weeds, condensation inside sealed equipment and insects attracted to the warmth of electrical equipment, all of which can be troublesome.

Many alarm companies require disclaimers on workability.

Police do not continue to respond to alarm systems

that lose their credibility by a high rate of false alarms, and most companies quickly learn better than to abuse police service by relaying alarm signals that come from outdoor installations to the police.

The alarm company that maintains its own patrol division can respond to outdoor systems and absorb the false alarms without abusing the police. However, outside burglar alarm systems are rarely workable and should only be installed when a way can be found to control false alarm factors and ensure a system with at least a fair degree of reliability.

Sometimes effective outside protection is obtained by combining good lighting with an alarm system that activates a five-minute, self-resetting bell. This will often drive an intruder off without bringing the alarm company or police into the picture. Obviously, consideration must be given to keeping this alarm from becoming a nuisance to people living nearby.

Point-to-Point Microwave

A recent development known as the **Point-to-Point Microwave Unit** has shown the potential to solve the outside protection problem for those whose greater need for security justifies its greater expense.

The device uses microwave energy between a sending point and a receiving point in a manner similar to the 'laser beam' and photoelectric cell. But, instead of a linear beam, the microwave energy field fills an adjustable volume up to six feet high and 25 feet wide with a maximum length of 1,000 feet per unit. The field must be broken over a substantial portion of its cross-sectional area to start an alarm. A small child would do it but wind, rain, birds and small blowing objects would not.

A test of this unit in an outside installation beside an existing 'laser beam' system showed this dramatically. In six months operation, the microwave unit worked without fail or false alarm while wind and rain frequently triggered the 'laser beam' system. The false alarm rate for a growing number of such installations remains near zero.

Two newer devices have also been introduced which, though they are not yet thoroughly proven in all working situations, appear to hold great potential for the protection of outdoor areas.

The first, called **Electronic Field Fence,** uses a special two-wire fence which can be installed on top of a roof or ordinary chain link fence or as a second perimeter inside a chain link fence. One wire, like a transmitting antenna, carries a 10,000-cycle signal from a generator and broadcasts an elliptical field to the other wire. The receiving wire actually monitors the signal for any change caused by an approaching object grounding out the field. The sensing wire goes into alarm at a preset variation that would be caused by a human intruder but not a small animal or blowing object.

The second device, the **Linear Microphone** is a small coaxial cable which picks up the 'sound' of tiny vibrations through changes in a static charge in the insulation between its two conductors. A separate unit, much like a hi-fi amplifier in appearance, measures the static field in up to 1,000 feet of cable. It is programmed to pick out those sounds which bear the characteristic 'signature' of an intrusion. Properly set, it should detect attempts to climb, cut, burn, or jack up a fence. It can also be set to count occurrences of a particular sound and alarm only after a specific number to screen out random tapping on the fence by children or small animals.

Other exterior devices such as ground vibration detectors or mercury tip switches on fences are available but not normally applicable unless on-premise guards are able to monitor the areas.

All of these devices tend to be most effective in controlled areas where standing guards are already nearby—available to investigate alarms before a more complex response from off the premises is called in.

3. Holdup Alarm Systems

Beware: Holdup alarms can precipitate violence.

A holdup alarm system may be added to any burglary system or may be an independent system. When added to a burglary system, an additional separate transmitter is always used to differentiate the holdup signal from a burglary signal at the central station. Police response to a holdup alarm is as fast as possible because of the threat to life. Hence, the police must know in advance that this is the nature of the alarm.

A note on safety: Two thoughts must be foremost in the installation and use of holdup alarm systems: First, attempting to use the alarm can create a danger to life; second, false alarms cannot be tolerated.

Holdup buttons are commonly located under a counter where employees face the public, but may also be placed elsewhere, such as in the walk-in box of a liquor store or in a vault, on the common chance persons may be locked in these rooms in the event of a holdup.

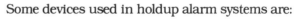

Some devices used in holdup alarm systems are:

Holdup Button

Holdup Button. A specially designed switch that can only be alarmed by pressing from two sides simultaneously or pressing a recessed button, thus preventing accidental alarms. This switch is usually reset with a key in order to verify the source of the signal later.

Foot Rail

Foot Rail. A rail about five inches above the floor in a protected enclosure which is raised by foot pressure to actuate the alarm. This device is a common source of false holdup alarms.

Cash Register Button. One of the ordinarily unused buttons on the cash register may be used as a switch that opens the cash drawer and sets off the alarm at the same time. Many cash register companies will prepare such a button, which is then connected to the holdup system by the alarm company.

Money Clip. A fixed clip in a cash drawer into which a bill is inserted. If the bill is removed, two contact points touch and cause a signal. While commonly used in cash registers in the past, it is not now recommended because the wear and tear on the wiring in a movable drawer (plus human error) makes this device another common source of false holdup alarms. It has wider acceptance in built-in cash drawers that do not have as much movement, such as those used in banks.

Money Clip

Remote Button. A wireless button that can be carried concealed on the person. Questionable reliability because the instrument is electronically sensitive and subject to battery failure is being overcome by design improvements. The money clip also comes in the remote format.

Remote Button

Holdup Lock. A special front door lock which, if turned one way, opens the door but, if turned the other way, opens the door and also sends a holdup signal. Another version uses two keys—one opens the door only, the other also sends an alarm. These are used when the danger exists of an owner being forced to reenter his premises when he leaves alone at a late hour. This device too has been a common source of false holdup alarms.

Motion Picture Cameras. A holdup system may use motion picture cameras which start filming when a

Camera

holdup alarm is actuated, thus supplying a filmed record of the event. This is useful in situations where the subsequent capture of the bandit might be worth extra effort. Videotape equipment is also used for this.

For retail businesses, perhaps the best holdup protection so far devised is the installation of large, obvious, one-way deposit safes to hold bills and change. Only a small amount of cash is kept in the register. Thus employees can be instructed to cooperate, and robbery is reduced to a much smaller event in terms of loss or danger to employees.

A note on safety: It is important to remember that silent holdup alarms MUST BE SILENT. Transmitters that emit clicking sounds, etc., must be located out of hearing range. The use of a holdup alarm must always follow careful consideration and planning. Because of the possible loss of life from the activation of an alarm in the presence of an armed robber, a commonly used option is to wait until the robber has left the scene before tripping the alarm. The rapid police response to a holdup alarm will often result in the apprehension of the robber even after this delay.

4. Transmitting Devices and Connecting Lines

A critical subject...

...too often left hazy in alarm planning.

The minimum alarm system discussed in the chapter called 'Basic Burglar Alarm Protection' was built upon the premise that any alarm signal initiated by a sensing device would be relayed within one minute to the central station. The system was connected via regular telephone line to the central station and sent a signal from an automatic telephone dialing device.

This basic alarm system can be substantially improved by upgrading the electronic capabilities of the transmitting device and the connecting line to the central station.

When deciding on
transmitting devices...

...beware of alarm
industry catchwords
that can mislead.

What you need to know
is not the name...

...but what it does.

Your choice in method of transmitting the signal to the central station will depend, to a large extent, upon the level of security your particular business requires and the exposure of your telephone line to tampering.

At present, six types of transmitting devices and the corresponding connecting lines are available for burglar alarm use. All six require the leasing of a line from the local telephone company. They are:

Digital Dialer and Ordinary Telephone Line

This type of transmission format has not traditionally qualified for UL Approval because the central station has no practical way of knowing whether the line is in working order. This deficiency led to the addition of a local bell in our example, so that even if the connection to the central station is broken (accidentally or deliberately) the premises are not without some form of protection. (If you want to know how easy disabling your phone line would be, go outside and inspect the phone connection to your premises.)

However, because of the generally high reliability of telephone lines and the low cost for this type of connection, its use as a backup system or with modern automated dialing equipment is becoming the accepted standard for minimum security.

The automated dialing equipment we refer to here should not be confused with its less reliable predecessor, generically referred to as a 'dialer,' or 'tape dialer' which phoned police with a spoken, tape-recorded message.

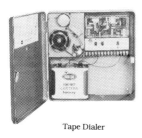

Tape Dialer

The newest generation of dialing equipment, called **Digital Dialer** or **Digital Communicator** is a highly reliable device that sends an electronic digital code to a receiving unit at the central station. The affordability of modern Digital Communicators

Digital Dialers ...

Inexpensive...

Versatile...

Reliable...

allows a reasonably high quality of security to be extended where before, without significantly increased costs, the only possibility was a local ringing bell. Today, a good Digital Dialer can handle eight separate alarm zones for burglary, fire and holdup.

A combination of a local bell and a Digital Communicator using voice-grade phone lines may now be considered as the best minimum alarm signaling system available.

Another desirable feature to the Digital Communicator is the immediacy of installation. Unlike other alarm lines, the Digital Communicator does not require special phone company engineering which can take up to a month. The Digital Communicator can be connected directly to an existing phone line in your building. Even if you are going to have a more secure line installed later, the temporary use of the Digital Communicator will carry you over the telephone company lead time.

Because of the potential widespread application of Digital Communicators, UL has conducted a two-year evaluation of this type of equipment and developed standards for its approval. There are presently units on the market that have UL approval. As an example of the features they must have to compensate for the lack of a direct link to the central station, UL requires the ability of the Digital Communicator to make repeated attempts to 'dial' the central station in case the first call does not go through. Many also have 'line-seizing' capabilities which allow them to override the normal traffic. Without this line-seizing capability, the Digital Communicator should not be used.

Today, many Digital Communicators can send 'opening' and 'closing' signals to the central station. One now on the market will identify up to eight different opening and closing codes for authorized people. The UL approved models must also send a signal to the central station from a built-in timer every 24 hours.

Take note that many Digital Communicators currently being marketed do not send regular signals to the Central Station when you close your business each night. Therefore, unless you specifically check the device regularly, you won't know whether it is able to send a signal or not. Supervising such devices is possible but not very practical. It is done by sending periodic signals from the Central Station to the protected premises. The response from the device determines the continuity of the telephone line. This can be programmed to take place as often as wanted. This is not usually done, however, for the simple reason of cost. Being connected to the regular telephone line, the Digital Communicator incurs a charge, just like a regular telephone call, every time a signal is sent. Even at a low cost per call, if the line-integrity function is used often, the monthly phone bill would reach beyond all the other costs of the security combined.

But difficult to supervise.

Digital Communicator

Where the telephone line is leased and paid for on a monthly basis by the alarm company without regard to traffic, the line security of the Digital Communicator can be extended to a very high level of reliability. The Digital Communicator with computer capabilities must not be seen as inferior. It's limits in this area are based on current company telephone rate structures. While the use of WATS lines is possible for carrying alarm information, they are considered inferior electronically to the basic voice line.

While the Digital Communicator is a fast-growing form of connection to the central station, the majority of alarm systems are connected to central receiving stations by circuits provided by the phone company especially for alarms. The most common of these is the McCulloh Loop Circuit.

McCulloh Loop Transmitters and Circuits

An industry standard...

The McCulloh Loop is a simple party line circuit designed especially for alarm systems that connects up to 45 subscribers to the central station via one or more telephone relay stations. This is the least expensive of all special alarm connecting lines because its one loop circuit can do multiple service for many customers while remaining free of all voice communications. It is very effective and has been used by most central stations as their main carrier of alarm circuits. Its use is now declining because of the introduction of the more versatile Multiplex Alarm Carrier and the growing acceptance of the less expensive Digital Communicators.

...now losing ground to digital dialer technology.

When a McCulloh Loop is used, it is important that alarm companies avoid overburdening this circuit with too many customers on one loop, a mistake commonly made in an effort to hold costs down.

The transmitting device traditionally attached to a McCulloh Loop Circuit sends an alarm signal by mechanically opening, closing and grounding the circuit in a coded pattern, like a telegraph key. Adding transmitters allows additional codes to distinguish between separate alarm circuits such as fire and burglary.

Today solid-state equipment is gradually replacing these telegraphic devices as the means of transmitting alarm signals. Some manufacturers now offer transmitters which can use an existing McCulloh Loop Circuit for the more complex information exchanges which will be discussed under Multiplexing. Multiplexing over McCulloh lines is very slow and is not in widespread use today, however.

Unlike the ordinary telephone line, a McCulloh Loop is designed to signal the central station when a short or ground occurs anywhere in the telephone circuit. Unfortunately, it is usually impossible for central station personnel to identify where a short or ground in the telephone line has occurred. This situation calls for an investigation by an emergency service department of the telephone company. The search can take up to 24 hours. While a ground will not necessarily incapacitate a McCulloh Loop Circuit, it can, leaving up to 45 subscribers unprotected during the delay.

At times when the phone company has many service problems, such as during major storms, a McCulloh Loop Circuit is as susceptible to interruption as any telephone circuit and usually cannot be restored as quickly because it cannot be rerouted to working lines.

A more serious weakness of the McCulloh Loop (as well as any unsecured phone line) is that a burglar can cut it before breaking in. The central station will receive a signal that something is wrong on that McCulloh Loop. However, the signal tells the central station only that there is a ground or a break in the circuit, not where the problem is. While the burglary is in progress, a telephone company service crew will be out driving from checkpoint to checkpoint over miles of city streets looking for a problem that could be anywhere on the loop.

Also, it is often easy for a burglar with some knowledge of electricity to remove an alarm circuit from a McCulloh Loop without causing any signal at the central station.

Unfortunately, precautions are seldom taken to secure these lines from tampering. Often, incoming lines terminate in an unlocked area of a building, open to the public. Sometimes the alarm circuit lines are even tagged separately so that phone company servicemen will recognize them and not tap into an alarm line (a common problem). The alarm company usually cannot do anything to make these lead-in lines more secure.

The McCulloh Loop Double Drop

Adds line security

In this type of connection, using the same type of McCulloh Loop transmission line, the McCulloh Loop is brought into the building at TWO locations at least ten feet apart and wired so that if one drop is cut, the transmitter is activated and the coded message is sent through the other drop to the central station. While hardly invincible, the Double Drop provides a beginning measure of line supervision to protect from tampering.

The Double Drop is not installable and is not necessary when the lines into the building are underground.

Direct Line

As the name implies, this is NOT a party line but is the subscriber's own private line directly to the central station

Becoming extinct...

Prior to the development of Multiplex, the only way to ensure alarm circuit and telephone line integrity with a high degree of reliability was with a direct line. It was, and still is, used in the protection of the highest-value locations.

But very reliable and highly respected.

Unlike McCulloh Loop, the Direct Line is not a passive system, waiting for something to happen to actuate its transmitter. It is an active system which originates at the central station where a current transmission device maintains a specific voltage over the telephone line, through all the wiring and devices of the alarm protection circuit and back to the central station where the voltage is constantly monitored for interruption or minute fluctuations.

The control device on the protected premises, commonly referred to as the Direct Line transmitter, primarily serves as a day/night switch and power control center for the alarm system.

When the building is in use, the alarm system is turned to a 'day' setting rather than off. This leaves most of the circuit still under central station supervision, while removing from the circuit those devices such as door switches which would continually activate signals. The Direct Line transmitter is sending voltage through the alarm wiring to monitor the system against tampering or damage during normal hours of business operation. Devices that stay in the 'day' circuit would include skylight lacing, foil on glass, power supply equipment and much of the alarm circuit.

A Direct Line is almost always a supervised system with a central station recording the time that the alarm circuit is switched from its day to its night mode of operation. Upon registering the change, the central station sends a specific signal back to the protected premises telling the subscriber it is aware he has turned the alarm on.

There is a possibility of someone duplicating the electrical profile that the central station is monitoring, and thereby bypassing the alarm system. To prevent this, an extra protective device which can be installed at the central station 'scrambles' the voltage pulse so that the current characteristics of the line cannot be duplicated by interceptive devices. This is called line supervision.

For a number of reasons, the Direct Line is becoming less frequently used. Because it is for the exclusive use of one subscriber and because it requires a pair of hard copper lines, the Direct Line is much more expensive than any other type of telephone connection.

Also, due to the scarcity of installed copper wire and the preoccupation of the telephone companies in establishing other and more sophisticated carrier systems, the Direct Line is becoming not only very expensive but also unavailable in some places.

Multiplex Alarm Carriers and Transponders

Gaining ground because of versatility and high security

Another type of connecting circuit is the Multiplex Alarm Carrier, perhaps the most sophisticated and electronically complex of all lines. Yet its practical applications are easy to grasp.

Essentially, Multiplexing provides two important improvements over any other method of alarm signal transmission: 1) it can carry answers to numerous questions about a protected site, and 2) it can do this with great speed.

Multiplex Transponder

Every good alarm system is reporting at all times that its circuit, including telephone line, either is or is not operating undisturbed. It is 'answering' this one question continuously. A Multiplex system is answering many such questions instead of just one. It does this by sending an electronic pattern composed of many questions to a transponder located at the protected site. This transponder replies to the questions, reporting on the condition of many circuits within the total alarm system. The condition of each is also displayed on the transponder at the protected location.

A few of these questions are:

- continuity of phone circuit
- condition of alarm On/Off
- condition of six to sixty-four alarm zones
- continuity of alarm wiring during day
- condition of power to alarm equipment
- condition of standby battery power
- total line resistance

Multiplex: not a device...

It's a technique...

In this way the central station is able to maintain a continuous survey of much information about conditions at the protected location. Each point of information in the total alarm system is scanned every few seconds, and with each scanning the central station receives replies pinpointing the location and nature of any disturbances or alarm conditions.

Thus many questions are answered about any emergency to forewarn police and alarm company service personnel as they respond to the alarm. This foreknowledge of conditions greatly increases the effectiveness of the response.

...for compressing information into microseconds...

...and splitting a wire into multiple transmission paths...

Multiplex systems are extremely flexible and can be designed with many variations to meet the most demanding security needs. If desired, the central station can use the Multiplex system to perform certain functions at the protected site remotely, such as turning switches off or on, locking or unlocking doors, sending coded instructions to patrolmen, etc. However, this requires special equipment which many central stations do not purchase.

...by dividing electrical frequencies.

Multiplex type signal processing does not require any additional protective devices such as 'double drop' or 'line supervision,' since its own inherent features exceed any that might be added. The Multiplex system is given UL's highest line rating (AA).

But beware: the term Multiplex is often misused.

The Multiplex connecting line provided by the phone company is very different from the McCulloh Loop circuit. It is of a higher grade, electronically 'cleaner' and less subject to resistance fluctuations or noise interference. The telephone line from each protected site is connected to an amplifier bridge at the telephone company where the signals are enhanced and combined with those from several other sites for transmission to the central station.

An advantage of the Multiplex circuit over the direct line is that it offers an economy based on group use. With many subscribers sharing the same Multiplex circuit, the connecting line charges for each drop are often not much higher than for a standard McCulloh Loop line.

Other features of Multiplex:

The Multiplex line is 'clash-free,' that is, alarm messages sent simultaneously by different subscribers cannot clash with one another or temporarily block each other as they do in many McCulloh Loop systems using older equipment. (ALL good McCulloh alarm systems, however, have Anti-Clash Transmitters to avoid this possibility.)

Multiplex systems are compatible with the most advanced computerized alarm systems and can provide fully automated alarm monitoring and supervision of openings and closings.

The telephone company can transmit Multiplex information over microwave and optical paths as well as hard wire lines. This means that Multiplex systems are compatible with advances being made in signal transmission.

Tampering with the circuitry of a Multiplex system (either within the alarm circuit or on the connecting line from the transponder to the central station) is extremely difficult. Attempts to tamper are instantly signalled to the central station along with the location of the disturbance.

Built into Multiplex equipment, and some other alarm equipment, is the ability to monitor the precise electrical resistance in the alarm circuit. This is referred to as **end-of-line resistance.** If a device or portion of the alarm circuit were taken out of the system during the day when the alarm is off, the alarm transmitter would show the alarm circuit in a trouble status the moment it happened, or in some cases when it is turned on. This is to guard against inside tampering with an alarm circuit.

Line Security

The reliability of the connecting line between the central station and the protected premises is a subject that often goes unmentioned in discussions between the alarm company and the subscriber.

In fact, this is a subject of vital importance and is one of the major distinctions between the different types of alarm protection.

Reliability of the communication line...

Electronic surveillance can be interrupted by any one of several causes: acts of nature, equipment failure, cut phone lines or more sophisticated attempts to compromise the line without actually interrupting service.

...and...

To provide ideal line security, a central station alarm company should know when there is any interruption in the telephone circuit providing alarm service, where the interruption has occurred so that it can be repaired and whether the interruption is an attempt at sabotage or just an accident.

...security of the communication line...

None of the four types of telephone connecting lines discussed provide this ideal line security. The strengths and weaknesses of the four types of lines can be summarized as follows:

...are different ideas.

With the voice grade line, service interruptions are relatively infrequent and usually bypassed by automatic switching. However, there is no monitoring of line continuity and cutting the line going into the building will disconnect the alarm from the central station.

The difference is very important.

With the McCulloh Loop, continuity of service is more limited, due to the effect of many subscribers on one party line. Monitoring of continuity is better in that a failure is usually detected at the central station but without any specific indication of where it occurred.

The Direct Line provides the highest possible level of line security because of the fact that when a problem occurs, there is no question about who is affected, as each line is a single subscriber, and when a single subscriber is identified, decisions become much simpler.

Line Fault Monitor

Multiplex equipment, because of its inherent capabilities, usually can provide this specific subscriber identification except when the telephone company is having a problem on the entire Multiplex circuit, in which case a number of subscribers will be without alarm service during the time it takes to put the circuit back into operation.

Neither the Direct Line nor the Multiplex circuit offers the self-correcting ability of the ordinary telephone line. For this reason, high security risks such as jewelry stores will often use a Digital Dialer as a backup system. A new type of Multiplex Transponder comes with a built-in Digital Dialer that only operates when the Multiplex line fails. This combination gives an excellent level of service.

Attempts to evade detection by duplicating the electronic profile of the alarm circuit, and thereby fooling the central station, can only be dealt with by a Direct Line with line supervision or a Multiplex Carrier.

Standby Power Supply

Usually standard

The alarm circuit must be supplied with standby power usually located in the transmitter. This power supply/transmitter links the telephone line, the alarm circuit and the building power source.

Power to the transmitter and all the alarm devices within the circuit is taken from the building power through a 6- to 24-volt DC transformer. This low-voltage is used to operate the circuit and to maintain a charge on batteries which will take over if the building power should fail. Thus, intermittent power failures or surges will not cause a failure of the system or even an alarm. A well designed system will send a signal to the central station indicating battery depletion of the main power supply.

Individual devices that use power, such as photo-cells, ultrasonics, microwaves, infrareds and capacitance devices, will draw power independently from the building through the closest outlet and have their own standby batteries. These outlets must never be turned off at night and the circuit breakers or wall switches controlling them should be tagged with that warning.

Radio Telemetry—The Future?

Don't expect your local central station...

In some newer systems, the telephone line may be eliminated by the use of radio transmission. Radio telemetry is the generic term being used for wireless transmission of security system signals. These radio signals are transmitted both within the security system from device to alarm transmitter and from the alarm transmitter to a decoding monitor in the central station.

...to have this.

Using UHF, VHF and FM frequencies, one decoding monitor can handle up to 10,000 transmitters from as far as 30 miles away or more using signal repeaters. The decoding monitor can also poll or interrogate each monitored location to ensure the system's integrity. The use of the airwaves for alarm transmission within a metropolitan area such as Los Angeles, however, may have insurmountable difficulties due to congestion by competing transmissions and interference from highrise structures.

5. Supervision

At several places through the book we have mentioned the ability of the central station to supervise activity of alarm equipment at the protected site.

Supervision can be vital to your security.

As supervision is somewhat intertwined with the subject of transmitters, connecting lines and line security, it is easy for this fundamental concept of electronic protection to become confused.

Therefore, we will restate some information from previous sections to help clarify the important distinction between alarm systems which are 'supervised' and those which are not.

Non-Supervised Systems

The minimum alarm system of our example is 'non-supervised.' That is, the owner of the business attends to the openings and closings himself, without supervision from the central station. The user

Least expensive and easiest...

turns the alarm system on with a key after he locks his doors at the end of the day and turns the alarm system off with a key when he arrives at the premises the next morning. A variation of this provides a 30- to 60-second time delay on the transmitter, allowing him to turn the system on and exit the building before the alarm system arms itself, and a corresponding time delay to allow entry in the morning. The system is on overnight while the premises are empty. The central station is not aware whether the system is on or off.

...but leaves you vulnerable

In this situation, the central station can receive alarm signals only during the hours that the system is left turned on by the subscriber. Therefore, any signal from this system received by central station is considered to be a burglary and is so treated unless proved otherwise.

There are several limitations to the effectiveness of such a non- supervised system. For one—a surprisingly common one—the subscriber forgets to use his key to turn the system on at closing time, or the employee who has this duty forgets. For a number of reasons, subscribers or their employees may not be reliably regular in turning the alarm system off and on as needed. Each failure to activate the system, of course, simply leaves the premises unprotected and completely at risk. With a non-supervised system, you could leave the alarm off for six months and the alarm company would never know it.

Another problem common both to non-supervised, and supervised, systems: the subscriber is responsible for operation of the system at all times of irregular entries and often forgets to do this correctly, thus causing an undue number of false alarms and danger to himself. Imagine what would happen if the owner of a business returns to his office at 10 o'clock at night to do some paperwork and forgets to turn off the alarm (non-supervised) or call the central station (supervised). The central station receives the signal and relays it to the police. The owner could find himself face-to-face with an armed policeman who thinks he is a burglar.

And a worse problem with non-supervised systems: the everpresent risk that unauthorized persons may obtain and copy the key.

To upgrade such a system we can change it from 'non-supervised' to 'supervised.' This simple change immediately gives a greater degree of security.

Supervised Systems

With a simple 'supervised' system, the owner or last employee to leave the premises at closing time closes all exits and tests the alarm. Upon getting a clear-to-set signal from the alarm system, usually a light on the transmitter (all non-supervised systems must also have this test light), the employee turns it on with an interior switch, and leaves. The opening of the door as he leaves activates the alarm system and sends a signal to the central station giving notice that the system is set for the night. After sending the 'closing' signal, the alarm system resets itself. When the door is opened at the start of business hours the next morning, another alarm signal goes to the central station. By comparing this signal to the prearranged starting time for that business, the central station knows that it is time for the system to be switched off for the day. The first employee to enter turns off the alarm switch.

Besides adding security...

...supervision gives you...

...a good management tool.

If entries are made during closed hours, the central station is notified by telephone so that the entering signals received are not interpreted as burglaries. (It is always best to make this call in advance.) A written record is kept of all openings, closings, and entries at non-business hours, specifying who entered, at what hours and how long they stayed. Of course, such entries are restricted to those with prior written authorization. Most alarm companies will provide a copy of this record weekly or monthly for a charge.

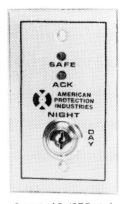

Supervised On/Off Control

The risk of forgetting to set the alarm is minimized by supervision, since the central station handles this routinely according to the subscriber's instructions. If the central station fails to receive the closing signal at the appointed time, an investigation can be made, usually within an hour.

The basic cost of adding supervision to an alarm system would be about $15 to $20 more per month to cover the additional central station monitoring.

Increasing Supervision

Supervision may be improved even more by changing the type of alarm transmitter and telephone line connection between the alarm system and the central station. A Direct Line or Multiplex system permits much more extensive supervision than is possible with a McCulloh Loop or a Digital Communicator.

Since a Direct Line or Multiplex system establishes two-way communication, the subscriber's signal to the central station that he is closing for the day brings a reply signal, usually lighting a light on the transmitter, assuring him that the central station has actually received his signal, his system is operating and his premises are now being monitored.

The Direct Line is also used for situations that require 24-hour monitoring, such as entries into computer rooms, tape storage areas, vaults, high-security zones, etc., all of which is also available with a Multiplex type of system.

Where high security is a must

Perhaps the most important advantage of the Direct Line and Multiplex system is that they make possible the installation and supervision of a dual system that monitors both day and night circuits. While the 'night' circuit is wired to provide an alarm at all entrances during night hours only, the 'day' circuit covers fixed windows, skylights, and ordinarily unused points of entry so that these may be monitored on a 24-hour basis. Any accidental damage to or tampering with sealed windows, skylights, alarm circuit wiring, etc., during the day (by a dishonest employee, for example, trying to prepare an entrance for night burglary) would send a signal to the central station. The subscriber would promptly be notified and a service vehicle sent to repair the system and restore it to service.

A note on police panels: Some communities permit Direct Line connections to police stations connecting to a panel owned and operated by a contracting alarm company. Please note that a 'Direct Line' in this instance is always non-supervised. Police stations are not equipped to give the kind of supervisory services that are provided by alarm companies. Usually response to these systems is very fast, being 'in-house' for the police.

No one system can cover everything.

The next degree of improvement in supervision of an alarm system is made possible by the versatility of a Multiplex alarm carrier system. With Multiplex, all of the previously described kinds of monitoring are possible plus a limitless variety of special and finely detailed modes of supervision. When a Multiplex Line is used in combination with another system, such as a Digital Communicator, two distinct telephone circuits are carrying signals to the central station, so a downed phone line will not leave you unprotected.

Design yours to meet your dangers.

Multiplex systems can be programmed, for example, to visually depict a floor plan of the protected premises on a screen at the central station, pinpointing precise locations of any irregularities on this diagram. However, this feature is not yet available in most alarm company central stations.

Each advancing degree of supervision gives greater control over internal security as well as greater protection against intrusion. Since the problem of theft by employees has grown so markedly in recent years, it has become advisable to use some degree of supervision in almost every alarm system.

In general, upgrading an alarm system by increasing the degree of supervision is an effective means of improvement and is especially desirable because the cost of increasing supervision from one degree to the next is relatively low.

6. UL Certification

Underwriters Laboratories, Inc. performs an excellent service for insurance companies, alarm companies, their clients and everyone concerned with security by establishing sets of Standards (virtually the only within the industry) that identify distinct levels of security. Thereby all parties have identifiable and labeled sets of conditions to use and refer to in making agreements to their mutual satisfaction. Without these useful classifications, discussions of alarm systems and negotiations involving security standards might sometimes be chaotic.

An insurance industry requirement for many businesses

Although it has been the preeminent agency that certified individual alarm systems, UL is no longer alone in this arena. Factory Mutual and Kemper Insurance, which are two companies that started in fire insurance certification, now also certify burglar alarm systems. At this time, very few alarm companies are familiar with their standards.

UL Certifications of individual alarm systems are strictly defined so that carriers and insured parties can form agreements on the degree of protection required by specific situations and both be assured that the standards upon which they have agreed will be similarly understood by themselves and alarm companies.

Alarm companies which are so authorized by UL can provide a UL Certificate for an individual alarm system verifying that it meets the levels and degrees of security specified.

Extent and Grade

Letter and number symbols are used to identify the degrees of certification. The letters designate the 'grade' of the system and the numerals designate the 'extent' of the system. Each system is first graded by two factors: the type of telephone line and transmitter used to connect the system with the central station and the guaranteed time limit within which the alarm company will respond to an alarm signal from the system. The 'grades' assigned are, in ascending order of degree, C,B,A, and CC, BB, and AA.

By 'extent' is meant the type and completeness of the alarm system equipment and installation, and the 'extent' designations, in ascending order, are 3,2, and 1. Thus the highest degree of certification is AA-1 while the lowest degree is C-3.

UL is presently discussing a grade AAA. This would require an extra secure method of transmission, and would surpass the requirements of AA.

The primary UL Certificate applies to supervised alarm systems connected to a UL Approved central station. Non-supervised alarm systems, such as those connected directly to a police station or using an ordinary telephone line (Digital Communicators), qualify for different UL Certificates. These certificates may bear the same grade and extent designations, but are distinguished by colors signifying a lower level of overall security.

A note on safes and vaults: Safes and vaults may be certified separately. The 'grade' designations with regard to safes and vaults remain the same but the 'extents', instead of being numbered, are designated either 'partial' or 'complete.' A vault certificate, for example, might read 'A-partial.'

Certification is ...

The 'partial' designation requires that the door to the safe or vault be contacted with a special UL Approved high-security contact switch. The 'complete' designation requires a door contact and protection of all four sides. In the case of a safe, this would usually be a capacitance or vibration alarm. In the case of a vault, this is normally a sound alarm within the vault.

...sometimes negotiable.

Although UL Certification provides many conveniences for everyone, those most concerned with choosing the degree of certification are usually the insured and the underwriter, whose responsibility it is to determine whether or not UL Certification is necessary for any particular carrier and, if so, what degree.

The bottom line is practical, reliable standards...

It is actually the alarm company that does the certifying of the system. UL approves individual alarm companies to issue UL Certificates on their own installations. To ensure that these companies maintain its standards, UL conducts testing and spot checks of installations.

...at higher cost.

Reference to the descriptions of UL Certification standards may be helpful in finding ways to upgrade any specific alarm system. The UL Specifications are listed in Appendix B.

A note on safes and vaults: Certification of alarm protection of safes and vaults is separate from UL's Certification of the safe or vault itself based on its construction for resistance to both burglary and fire. Where a safe or vault is required, be sure to ascertain the required ratings for insurance before purchase.

7. Auxiliary Devices and Services

Today many new services and devices are available which, although they are not part of the burglar alarm system itself, work hand-in-hand with the alarm system to extend the range and degree of security.

As a few brief suggestions of ways that security may be increased beyond the limits of a burglar alarm system, these may be of interest:

Closed-Circuit TV Monitoring

Wide price variations.

With advances in video technology lowering the cost and multiplying the applications, the use of CCTV to strengthen security is dramatically rising. A well designed CCTV system can allow a small security force to monitor a very large area and, combined with a video recorder, can keep visual records. A few

Shop around...

basic applications include loading docks, parking areas, high-security storage rooms and areas of high shoplifting potential. CCTV is also useful for the identification of personnel at entry points to sensitive areas. In extensive installations, the CCTV

...and weigh values carefully.

signal can be transmitted by microwave from the camera to a monitoring screen over distances up to ten miles at a suprisingly reasonable cost.

A new device called a **Telesentry Processor** can greatly simplify the problem of monitoring many cameras simultaneously. The telesentry processor monitors the view of several cameras at once by digital processing, comparing the images several times a second. When the image coming from any camera begins to change, it is queued up on a monitoring console being watched by a guard. With this system, fewer guards are needed to monitor many areas where activity is expected to be infrequent. Up to 20 cameras can be placed on one telesentry processor (which effectively makes each camera a motion detector) and more than one processor can be connected to a monitoring console. A video recorder can be connected into the circuit to record the queued image. An alarm reporting computer can also be queued to send an alarm signal to any specified destination upon the activation of any camera.

While this is an expensive system compared to standard alarm detection, the viewing capabilities can reduce guard personnel significantly and actually reduce cost in large installations. A 48-camera system installed with attendant equipment in one case was reported to have cost $95,000, but the yearly savings in guard personnel was $200,000, paying for the system in six months.

'Card-Access' Systems

Where keys are a problem

To control access to buildings or parking areas these cards, similar to plastic credit cards, can be revised and reissued at frequent intervals, thereby keeping tight control over all keys issued. A similar passkey system uses digital key stations which look like small hand calculators. A preset code must be keyed in to unlock the device. Card-access systems lend themselves to a wide variety of applications in controlling the access of personnel to sensitive areas.

While such applications complement central station security systems, they are not usually connected to a central station and so will not be discussed at length in this book. However, many alarm companies offer card-access installations as an extra service.

Motion Picture Camera Installations

Holdup situations

These keep a film record of entries or events in high-security areas, film holdups, etc. Cameras can be turned on manually or by a burglar alarm. While usually activated by holdup buttons, cameras can also be set into action by a burglar alarm device. These cameras usually give a 16mm recording of what transpires during the time they are activated.

Proprietary Alarm Systems

Many large companies keep a standing 24-hour guard force provided with a comprehensive surveillance system including conventional alarms, fire alarms, CCTV and inter-building communication

**Where possible, the
installing company
should also be the
servicing company.**

Often mandatory...

**...must comply with
municipal codes**

with a central console acting as a 'mini' central station on the premises. Usually, all fire alarms, and often burglar alarms, will be automatically retransmitted to an alarm company central station as well as indicated at the console. Such a system is usually designed by a security company and sold to the proprietor with a service agreement.

Emergency Notification Systems

Usually considered essential in highrise construction, these systems notify all or any part of a building's tenants of fire, bomb threats, riotous intrusions or other dangers to security. Though primarily designed for voice-directed evacuation in highrise buildings, these systems can be combined with background music, paging systems, elevator recall, fire door release, air conditioner shutdown, smoke exhaust fan activation and other functions.

Generally, an emergency notification system creates the possibility for tone and voice communication between the security center of a building and each floor of the building (or other assigned area). In an emergency, the system is automatically activated by a fire alarm condition and can send an audible tone or voice message to any floor or the entire building. A security officer at the main console can turn on the system and send a prerecorded message selected from a library covering any emergency. At any time, the console operators can override the recorded message to give direct instructions to selected floors in the building. The main console has its own emergency power and fail-safe devices to inform it when the circuit to any floor has lost the capacity to communicate.

Note: Emergency notification systems and proprietary alarm systems often fall under strict local or state regulations. For this reason, the selection of a company to install such a system should be made carefully.

8. Patrol and Guard Services

Patrol services provide personal attention in addition to (or sometimes instead of) the electronic surveillance of the basic alarm system. Work that cannot be done by electronic devices but requires personal human labor is naturally much more costly; however, upgrading the protection of any property by merely adding one or two patrol checks during the night is fairly inexpensive, usually $20 to $35 per month.

When this same patrol service can be used to respond to alarm signals that are received from the premises, security is greatly enhanced for very little more expense. Some full-service alarm companies have their own patrol service which, in addition to normal patrol duties, can be radio-dispatched to respond to alarm signals. Unlike alarm service personnel, who respond to restore alarm service, the patrol personnel can actually act as a private police agency. Where a central station alarm company does not have a patrol division, it often works cooperatively with the local patrol companies within its area of service. If an alarm company cannot provide this service, a patrol company can be contracted independently to act as your agent in responding to alarms as well as patrolling. The alarm company can be instructed to notify the patrol company upon the receipt of any alarm signal.

Usually standing guards are not contracted for by alarm companies. There are guard companies with which you would deal directly should you feel your security would benefit by having a guard on duty.

Many security systems could be improved simply by adding the personal attention of patrolmen who make regular route checks hourly, or at any interval desired. In some instances, no more than one nightly check is needed to assure that the building is properly locked, the alarm system is properly set, and that all has been left well for the night.

If you need guard service...

...make a point to get to know your guards and their supervisors.

Patrolmen and guards are commonly used to make 'key tours' of more complex premises, such as industrial sites, large shopping malls, highrise buildings, etc. By means of a 'Watchman Clock Tour' a complete and on-time inspection tour is provided. There are two variations. In one the guard uses a key at each station to send an 'all's well' signal to the central station, and each station must be visited at a precise time and in a precise sequence or the signals will not be accepted as proper by the central station. In the other, the guard merely carries a punch card clock from station to station punching a code with a different key located at each station. These punch cards are later reviewed by supervisors to ensure that the tour was made on time.

The minimum patrol service available is usually one nightly visit. The minimum guard service available is usually a four-hour shift. Guards may be engaged for regular duty shifts or on a one-time basis for special events.

The vast extent to which guards may be employed to strengthen a security system is beyond description here. It is obvious that each additional man-hour of personal attention to a property increases the security of that property greatly.

If you are considering hiring guards, ask yourself first if it is necessary for them to be armed. Most of the work guards do can be done without weapons. Not having the recourse to lethal force, a guard will usually follow other, more practical, options.

The liabilities and dangers of using armed guards cannot be ignored. Remember, you get what you pay for. A guard company charging $8 to $12 per hour per man may be only paying $4 to $6 per hour to that man. There is often little screening of the honesty and capability of guards. You are usually liable for their mistakes.

That concludes 'The Tools of Security.' Our next chapter will deal with fire and hazard alarms, primarily with an emphasis on the commercial users. Anyone concerned only with the protection of a home might move on to the 'Home Alarm' chapter. Take heed, though, that in the home fire is probably a greater threat to human life and property than burglary or intrusion.

FIRE, HAZARD AND WATERFLOW ALARM SYSTEMS

Any time you go home at night and feel concerned that something might happen at your business, you probably have a problem that can be solved with an alarm. No matter what the hazard, there is usually a way to watch for those conditions that would cause it, and, should they occur, trigger a prearranged response. The cost for this valuable service is relatively low.

Your business can survive a burglary.

The most common hazard, of course, is fire. But many other kinds of hazards are present in different businesses or industries. These can be temperature changes, the escape of toxic gas, loss of pressure in sealed liquid systems, drop of water level in a tank, or the malfunction of any continuously operating machinery.

Can it survive a fire?

In many cases, there is already some type of monitoring built into a manufacturing process or commercial operation and all that needs to be done is tie that into a transmitting device to extend the monitoring to a responsible authority.

In the protection against fire, you will find that your

decisions are simpler than in burglar alarm protection because the options have largely been defined already by approving agencies and government fire regulations.

Fire alarms work.

Compliance with these codes and regulations is, therefore, of major importance. Thousands of dollars of manpower and equipment can be wasted when a fire protection system fails to gain the approval of the responsible agency. It is common for the approving agency to ask for some small modification or the addition of a few more sensors. But a basic error, such as failing to put wiring in conduit where required, could lead to the necessity for the complete reinstallation of the alarm.

Our goal in this chapter is to give a basic understanding of the elements of fire and hazard protection, to help in those areas where choices are necessary on your part, and provide a basis for dealing with alarm companies and rating authorities.

If your main concern is fire protection, and you haven't read all the way through the book, we recommend that you first read the section called 'The Central Station.' The information there provides the foundation for understanding what we will be saying in this section.

How Does a Fire Alarm System Work?

Any central station system for the detection of fire or other commercial and industrial hazards operates on the same basic principle as the burglar alarm system.

The cost is probably less than you think.

The sensing device (trip-point) installed in a key location in a building sends a message to an electrical transmitter somewhere else in the building when a specific condition activates it. The transmitter then relays the message back to the central station over a telephone company line.

The hazard alarm system can be hooked to the central station by the same methods as the burglar alarm—either Direct Line, McCulloh Loop, Multiplex carrier, or voice grade telephone line. Since the probability that the line will be intentionally destroyed is not as great as in burglary protection, there is less justification for the extra cost of a Direct Line or the Double Drop often used in burglary systems.

Basic Equipment that is Needed

Because a different response is required at the central station, each alarm system installed, whether for fire or other hazards, requires its own sensing devices to detect a hazard and its own transmitter to signal the central station.

Fire alarm protection has to be done by the book.

To avoid any possibility of confusion between fire and burglary, most responsible alarm companies require a separate phone circuit for fire when a McCulloh Loop is used.

The use of a Multiplex type line and transmitter has the advantages of allowing all the systems to be combined, reducing cost of installation when multiple systems are needed, as in high rise buildings, and providing a constant monitor on the condition of each system.

The key is getting a qualified company.

In addition to a separate transmitter, two special pieces of equipment are specifically required by fire rating authorities for approval of a fire protection system:

Power Supply/Transmitter. Power failure often occurs with fires and can also be the cause of other industrial hazards such as equipment malfunction. The fire and hazard alarm system, therefore, needs to have its own source of emergency power or it can fail to operate due to the very condition that it should be reporting.

Approving agencies such as...

The Power Supply/Transmitter, which remains in a standby mode as long as normal power is functioning, can carry the system through blackouts, brownouts or power shortages for several hours. This device will also monitor each component of the alarm system, sending a trouble signal to the central station if there is a breakdown. It is required for an approved system and reputable companies will not install a system without one.

...fire marshall...

Signal Retransmission Unit. While many signals, for example one indicating that a machine has turned off, call only for the dispatching of an alarm company vehicle or notification of the building owner, any signal that indicates the presence of a fire must be relayed immediately to the local fire station. This is accomplished by an approved device known as the Signal Retransmission Unit, located in the central station.

...city building inspector...

Some fire departments have receiving units in their fire stations (usually installed and maintained by an alarm company) to receive signals directly from alarm devices on the protected premises. But in most cases 'direct' fire alarms are actually direct to the central station and retransmitted to the fire station. The alarm signal is prearranged with the fire station for each subscriber and an alarm received is always immediately retransmitted and then verified by telephone. The Retransmission Unit is tested several times daily to ensure its operation.

...or rating authority...

...are almost always involved.

Again, in all UL Approved central stations, a record is kept of every signal received and, in the event of an alarm, an instruction file is called to the operator's attention telling him what special procedures apply.

Rating Fire and Hazard Alarms

While insurance companies often require the purchase of an approved alarm system before writing a

burglary insurance policy, the general practice with fire insurance is to grant substantial premium reductions for properly installed and approved fire alarm protection. For setting insurance rates and discounts, the insurance industry relies upon several fire rating bureaus. Working in cooperation with Underwriters Laboratories, these rating bureaus set industry-wide premium rates and inspect individual alarm systems to determine appropriate premium reductions.

Rating Authorities

In the United States, the National Fire Protection Association (NFPA) Standards are the reference used by a number of rating authorities, in addition to Underwriters Laboratories, that certify fire and industrial hazard alarms.

One of the major fire rating bureaus is Insurance Services Office (ISO). When contacted by an insurance underwriter, ISO will inspect a business and recommend an insurance rate depending on the building's construction, location, contents and existing alarm systems. ISO will also inspect a new installation of fire alarm equipment and require any corrections needed to bring the system up to its standard before approving an insurance reduction. Monthly, semi-monthly or annual inspections and tests are performed by the alarm company, depending on the system, and kept for inspection by ISO.

In addition to ISO, which is primarily a fire insurance rating authority, other reputable organizations also rate fire alarm systems as part of their field inspections for all types of loss prevention. Three of the prominent companies among these are Factory Mutual (FM), Industrial Risk Insurers, formerly (FIA), and Kemper Insurance. Also, UL has recently entered the field of inspecting and certifying fire and water sprinkler supervisory systems.

Whatever the rating agency, the same rigorous standards applied by the Underwriters Laboratories, Inc. to the central station for burglar alarms are also required for fire alarms. ISO, for example, will approve discounts only for those systems connected to UL Approved central stations.

Minimum standards for any fire alarm system include a separate transmitter or transmission zone for each type of signal to be sent, a well tested procedure for retransmission from the central station to the local fire department and the installation of an adequate number of sensing devices, as will be discussed later.

What is Effective Fire Protection?

In fire protection, alarms are not enough...

Most commercial businesses are already provided some kind of protection against fire with equipment prescribed by local regulations. This can include fire extinguishers, special doors and signs and in some cases an automatic sprinkler system activated by heat and capable of dousing fires before they can spread.

Someone must take responsibility...

Where such a water sprinkler system is installed the fire regulations require a local alarm bell to be attached to the outside of the protected building to announce the activation of the sprinkler system and help arriving firemen locate the fire. But unless someone hears and understands the ringing, the fire department will never know that a sprinkler system has been activated, and, as sprinkler systems do not turn off automatically, water will continue to flow from all heads that have activated until manually shut off.

...to analyze the dangers and eliminate them.

Therefore, almost without exception, no building is adequately protected against fire without a 24-hour system for monitoring conditions within the building from a remote location, such as the central station. Such a system generally provides three methods of sending an alarm signal.

Three Means of Protection

The simplest, and often most immediate alarm is a direct message sent by people occupying a building. Therefore, most alarm systems begin with manual alarm stations strategically placed for quick response.

Unlike the burglar, who has to break in...

However, to protect against fire in unsupervised areas or at times when the building is unoccupied, a second system must be able to automatically detect and report the conditions that usually indicate the presence of a fire.

...the causes of fire are already there.

The need for a third type of fire protection arises where an automatic sprinkler system has been installed. The main function of this system is to prevent water damage. When activated, each sprinkler head on the automatic sprinkler system discharges water at a rate of as much as 50 gallons per minute—a rate that would fill an average swimming pool overnight. Just one sprinkler head set off in an office, perhaps by a cigarette butt tossed in a trash can as you leave for the night, could build up several feet of water behind your door waiting to surge around your feet by morning. Obviously, without some form of central station supervision, sprinklers could cause enormous damage to property by running unnoticed overnight or all weekend.

To find them...

...get professional help from fire department, insurance agent, alarm company.

The sprinkler system is part of the building's plumbing system and is not installed or maintained by a central station alarm company. However, alarm companies do install and supervise devices for reporting immediately whenever water begins to flow through a sprinkler head.

In practice, fire protection does not always employ all forms of protection. It is usually considered enough to have either automatic fire detection devices or monitoring of a water sprinkler system (in which case the water sprinkler heads become fire detecting devices).

Manual stations are almost always used in conjunction with an automatic fire alarm. They are also often used in conjunction with water sprinkler supervisory systems. Sometimes they are used as the only form of fire protection.

In the remainder of this chapter we will discuss in detail these three basic approaches to automatic fire protection under the headings 1) Manual Alarms, 2) Automatic Alarm Devices and 3) the Water Sprinkler Supervisory System:

1. Manual Alarm Stations

Manual Pull Station

These alarms, located in small red boxes, make it possible for anyone in a building who sees a fire that has not yet activated the automatic sensing equipment (if any exists) to send an immediate alarm signal to the central station.

There are two common UL Approved types. The **Manual Pull Station** is set off by a pull handle on the box. The **Manual Break-Glass Station** contains a small glass rod that must be broken or a window that is broken to reach a lever behind it. Several variations of each type are in common use.

Manual alarm stations are usually located in halls and at exits. When activated, the manual alarm station rings a bell locally and, if connected to the central station, brings emergency response from the fire department. Some current models are 'self-contained' in that they have a power supply and alarm transmitter built in for ease of installation. In this case, the building power and phone company line connect directly to the device.

A note on pull stations: There could be a temptation to underestimate the value of the manual pull station because telephones are so readily available to call the fire department. The prearranged, direct response set in motion by tripping of a manual pull station can save minutes over the more complicated approach of the average phone call. When there is a fire, those minutes can be quite important. Ask your fire department.

2. Automatic Fire Alarm Systems

Automatic systems to detect the presence of fire are far less expensive than water sprinkler systems and are generally used in buildings that do not have complete sprinkler coverage. The fast response achieved from fire departments makes this kind of system a practical and economical, though probably **Once installed, fire** incomplete, protection for any business without **alarms need little of** existing protection.
your attention.

If you already have an automatic sprinkler system, it may not seem worthwhile to install an automatic fire detection system. But insurance companies usually allow substantial discounts for the additional system because a properly installed automatic fire alarm will go off several minutes ahead of a water sprinkler head, often while the fire is still in a smoldering stage and has not yet reached its flash point. Even after fire reaches a water sprinkler head, it takes 60 to 90 seconds to activate the head (this delay may soon be cut to under 30 seconds). A minute at this stage is critical. When the combustible gases released early in the fire cycle eventually explode the fire escalates to major propor-
They are maintained tions instantly.
and tested by the alarm
company.
Upon request, your local fire department will usually send someone to your building to explain your particular fire problems.

Since all alarm signals from any one of the sensing devices used in the automatic fire alarm have exactly the same meaning as an alarm from a manual alarm station...that there is a fire...these two systems can be combined on a single transmitter.

There are three types of sensors in common use: **Heat Sensors, Smoke Sensors** and **Flame Sensors.**

Heat Sensors

The most common trip-point for early fire detection is the heat sensor which can transmit an alarm when certain changes in room temperature occur. There are two types:

Heat Detector

Don't assume wiring will be hidden...

Fixed-Temperature Heat Detectors normally trigger at 136 degrees with higher settings possible for hot areas such as over boilers.

Rate-of-Rise/Fixed-Temperature Heat Detectors are more often used. This device, usually contained in a small round case, will alarm whenever the temperature rises 15 degrees in a minute and also, in the case of a slow rise, at a preset temperature. Since the combination Rate-of-Rise/Fixed-Temperature Sensor costs only a little more to install, performs better and yields a higher insurance credit, many quality alarm companies do not install the simpler type except when it may fit into a special

...If it is important, ask. plan.

Usually an approved system must have at least one heat sensor in every room a person can stand up in, including closets. Sensors must be placed both above and below false ceilings.

Heat Detectors

On large, flat areas the devices can often be spaced up to 50 feet apart (a 25-foot radius for each) but must be closer together when obstructions such as wooden beams separate them. Beams of up to 2 by 10 inches can require as little as 15-foot instead of

50-foot spacing. A heat sensor must also be located in every skylight, in stairwells, under outdoor overhangs (a common way industrial buildings burn down is by fires started in trash bins alongside the buildings). These are just a sampling of the detailed specifications published by the rating bureaus for the placement of heat sensing devices under all types of construction that may be encountered.

Smoke Detectors

To detect a fire before it reaches its flash point, more sensitive devices are required. Because these devices are more complicated than the basic heat sensor, they are more expensive to install. But in most cases they are capable of detecting the presence of fire much sooner. They are always used in specialized areas such as computer rooms and vaults, and are highly recommended throughout. The types in use are:

Heat/Smoke Detector

Ionization Smoke Detector. Ionized particles of combustion which circulate in the air even before smoke is visible change the balance of an electrical circuit in this unit completing a circuit to the transmitter. Set at high sensitivity levels, these devices can detect overheating of the shielding of electrical wire and are thus useful in computer rooms and clean environments. The Ionization Detector is also used in attics, stairwells and halls of multistory buildings.

The ionization detector is very sensitive to the hydrocarbons found in cleaning agents and will false alarm when they are present in the atmosphere.

A note on radiation: This device contains a small amount of radioactive material. Care should always be taken to avoid prolonged close exposure.

Photoelectric Smoke Detector. Reacts to smoke passing through a photoelectric cell. Because of its

Heat/Smoke Detector

ability to visually detect some types of smoke which may not give off ionized particles, the photoelectric device has applications not suitable for any other type of detector. With proper engineering and servicing it is very reliable.

Alarm company engineers have long debated the relative merits of ionization and photoelectric smoke detectors. At one time the ionization method was favored and even today contractors may specify this type in their plans. However, due to improvements in the photoelectric type, principally the use of light-emitting diodes as a light source, they are now considered superior by most alarm engineers. Both require periodic maintenance and cleaning, preferably every six months.

In some situations it is desirable to use heat and smoke detection in the same area and units are manufactured which will perform both functions.

Flame Detectors

For highly flammable materials and outside areas

Flame Detector. This type of sensor visually searches out the flame with a principle similar to infrared detectors. An outside area with highly flammable materials is a good example of where the flame detector is of special value. Unaffected by ambient weather conditions, the flame sensor is highly sensitive to the emissions of infrared light given off by fire and has a wide field of vision. It can effectively view a large area, responding only to flame.

A note on fire regulations: Any building that is used for human habitation, such as a hotel, motel, apartment, hospital or rest home, falls under city, county, state and federal fire regulations covering alarm systems. Where threat to human life is a consideration, the government's regulations are very different and often much more demanding than those of the rating agencies. It is always a

good idea to contact your local fire department for advice before installing a new fire alarm.

3. Water Sprinkler Supervisory Systems

If you have a sprinkler system, read this!

The usual automatic sprinkler system consists of sprinkler heads spread over the ceiling of each room of the protected building. These connect to the main water supply by pipes called risers that come up inside or outside of the walls. Water sprinkler heads are located virtually everywhere that a heat detector would be located in an automatic fire alarm.

Special sensing devices are designed to detect water flow anywhere in the system and to monitor the on/off position of valves that feed the system.

Waterflow

The issues here are notification of fire department and water damage.

Waterflow Vane Switch

Proper supervision of the system requires a water-flow sensing device to be placed on every riser feeding the sprinkler heads. One such device, called the **Waterflow Vane Switch,** is usually a paddle-like feeler that can be inserted through a hole drilled in the pipe and sealed from outside. When water flows in the pipe, the feeler bends, causing a signal to the waterflow transmitter and then to the central station. These devices also come in self-contained models. A waterflow signal received at the central station always requires emergency responses by the fire department.

A note on freezers: In some high-risk areas or where sprinklers are installed inside a freezer, a dry system is used, holding the water behind a valve that is kept shut by air pressure in the pipe. These systems are also monitored for air pressure so that a leak in the system can be detected and repaired before the air pressure falls low enough to allow the water valve to open, possibly causing serious damage.

Valves

Included in the water sprinkler system is a series of manual valves located either inside or outside the building for use in controlling water flow and repair and maintenance. Seldom do companies have programs for regularly checking that these valves are open though the possibility of their being left closed through error or vandalism is always present. Some companies padlock the valves in the open position to ensure the sytem will operate when needed. When it is desirable to lock these valves open, it can be done with 'breakaway locks' which work normally but can be easily broken by a hammer blow in an emergency. Your fire department can usually direct you to where to buy breakaway locks.

At 50 gallons of water per minute...

...someone must know where the valve is.

And when to shut it off.

Central station supervision can monitor the valves constantly to ensure that they are not tampered with or left unintentionally closed. The device for doing this is usually a specially adapted mechanical switch attached to the valve in such a way that the valve cannot be opened or closed without activating the switch. This can be a self-contained power supply and transmitting device, or can be wired to a central power supply and transmitter in combination with other switches.

Valve Monitoring Switch

There are three types of valves in common use, each requiring its own type of switch:

The Outside Stem And Yoke Valve (OS&Y) is operated by turning a hand-sized wheel and is usually located on a riser about five feet from the ground.

The Post Indicator Valve (PIV) is literally a red post sticking three or four feet out of the ground. A handle on top of the post or a crank on the side opens and closes the valve. Its position is shown by an indicator behind a glass window.

The Butterfly Valve is identified by a crank-type handle usually located on a riser.

There is always a valve located at the connection to the city water supply and usually found in the 'vault' or 'pit,' a chamber in the ground near the property line.

A signal at the central station indicating that a valve is in the wrong position requires only that the subscriber be notified and a serviceman dispatched to investigate and correct the problem.

Auxiliary Devices

In some situations holding tanks and pumping equipment are included in the fire protection system to feed sprinkler devices. These elements of the sprinkler system must be supervised with special devices. The most common are:

Easily overlooked...

Water Level Supervisory Switch: monitors the level of water in gravity or pressure tanks usually found on the roof of a building. If there is not enough water, the sprinkler system may not be able to contain a fire. Too much water can damage a system or cause overflow.

Be sure the detection system is complete.

Fire Pump Supervisory Device: monitors the supply of power to a water pump feeding water to a firefighting system. This pump is designed to go on automatically if water begins flowing in the sprinkler system to ensure enough water pressure to keep the sprinkler system operating.

Air Pressure Supervisory Device: detects problems with air pressure in a dry pipe system (often used to run sprinkler systems through freezers).

All devices used in a water sprinkler supervisory system, as well as the installation itself, usually must be approved by UL or Factory Mutual. For approval of the supervisory system, a regular program of testing and inspection is required. This is most often monthly or bimonthly. Each valve monitoring device in the system is tested all the way through to the central station. The water-flow devices are tested by running water out of a special valve at the end of the system until a signal is generated and received at the central station. Records are kept of these testings and are periodically inspected by the rating authority.

Fire Insurance Rate Discounts

A subscriber to a fire protection system is not required to have it rated, but UL, ISO or one of the other rating bureaus will inspect any system and recommend substantial discounts for those it approves.

Approved fire alarm systems lower insurance rates...

Generally, the procedure is that the rating bureau sets a base rate for the particular insurance client, after inspecting his property, and then reduces that basic rate by specific percentage for every approved measure taken to protect against fire or water damage. Many rating bureaus apply the discounts separately for water sprinkler, fire alarm and manual alarm systems.

...often enough to pay for themselves.

To get a picture of what this means to you, ask your insurance carrier what your basic fire insurance rate is and then ask what the adjusted rate would be if you added any of the following: A) water sprinkler system, B) water sprinkler supervisory system, C) valve monitoring system, D) automatic fire alarm system, and E) manual alarm system. You might also ask what other suggestions your insurance carrier may have that would further reduce your rate. Compare these savings to bids from a local approved alarm company.

Discounts vary according to the quality of the equipment used and specific conditions within a building.

But the fire rating bureau will not approve any discount unless approved devices are installed and the system is complete.

There are several variations to the basic automatic fire alarm system that may be required in some cases by fire regulations and may or may not improve the fire insurance credit. Some added systems or adaptations of the basic system can:

- Require individual zoning of the fire alarm system (usually requiring approval of the fire department).

- Ring bells on the premises of a business or habitation (one per zone and usually mandatory).

- Close doors to prevent the spread of flames.

- Trigger the discharge of carbon dioxide, halon gas, or water for extinguishing fires.

- Shut down air moving equipment to retard flame spread.

In addition, some building construction may be necessary, for example, isolating stairwells, fireproofing walls, installing extinguishers and hoses.

A note on sprinkler costs: In taking costs into consideration, keep in mind the fact that water sprinkler systems require major plumbing installations and therefore cost 10 to 100 times more than automatic fire alarm systems.

Industrial Hazard Alarms

In industry, many possibilities exist for potentially dangerous and damaging equipment failures which call for immediate notification of the proper authorities. An alarm system can be designed for any situation.

Don't ignore a potential danger.

Failure in the freezing unit of a frozen food warehouse which is closed nights and weekends could result in a total loss of inventory. But a Temperature Control Switch set to trigger a supervisory transmitter at a predetermined temperature would allow the owner to be notified immediately of the danger.

Monitoring for it may be easy.

Devices called **Reverse-Thermal Detectors** are used to detect sudden decreases in temperature around liquefied natural gas, butane and propane containers. If a tank should leak, the expanding gas causes freezing temperatures, activating the alarm.

Water detectors can be placed in basement areas to detect sump pump failure, surface water or broken pipes. The detection of water is done both with mechanical and now electrical devices.

Similar industrial alarms are in use for water tank levels, factory furnace temperature, gas content and many more situations. Soil can even be monitored to detect harmful moisture or chemical changes.

Combined Alarm Installations

Most companies have need for both burglary and fire protection systems in addition to any other alarm systems that may be applicable. Once one system has been installed a second system can usually be added for less than the cost of an original installation.

Obviously, the most economical way is to install all the alarm systems that a business will need at once so that the time spent in planning and installing equipment and the interruption of business operation will not have to be repeated.

For that reason, many alarm companies, when they receive a request for an estimate on the cost of a burglar alarm system, make a practice of also estimating the cost of combining other desirable services such as fire and hazard protection systems.

If your thoughts are primarily focused on burglary protection when you ask for an alarm system estimate, take the time at this crucial point to examine any suggestion your alarm company may have to extend that protection. Your property, your business and your safety or the safety of your employees may be just as much at risk from less dramatic hazards.

Take the opportunity to examine your own property with an eye for hidden hazards that the alarm company representative may not see.

We would like to leave you with the thought that almost anything that could go wrong can be watched for electronically, usually at a reasonable cost. There is no cheaper time to take care of it than when you first contract for alarm services.

HOME ALARMS

Electronic burglar alarm systems installed in homes work upon the same principles we have already discussed. They must be thought out along different lines, however, because the values behind the protection are different.

First, the protection of the home involves things that are dear to you—your family and personal belongings. Second, the aesthetics of your home present difficulties in the installation which are not normally a concern in business applications. And finally, your finances are usually a limiting factor.

So, quickly, we will take a fresh look at the tools of security in light of your personal priorities, re-examining the assumptions which in earlier chapters formed the foundation of our judgements.

To begin with, you will need...

A New Point of Focus—Your Family

From our point of view, the ideal home alarm is one that will be working late at night when you are asleep in bed, anytime you or your spouse and children are home alone, and also when no one is home.

**Fire, the deadliest
threat...the easiest
addressed**

In contrast to our earlier emphasis on burglary, we believe the primary purpose of a home alarm, and one of the easiest goals to achieve, is the protection of a family from fire.

Again, departing from our premise that burglary protection is built around the concept of notification of authorities when an intrusion is detected, our goal in home alarms is most often creating a local deterrent—setting off bells, buzzers, lights or whatever it takes—to drive an intruder from your property before he can cause harm.

This deterrent usually begins with something as simple as an alarm company insignia visible from outside your home to notify the burglar the house is protected.

Deterrents work.

Generally speaking, the burglar who works in residential areas plans to make an easy entry, work undisturbed and leave undetected. He knows that if attention is drawn to him, he is highly visible. In a quiet residential neighborhood, anything that creates noise draws attention. He also is aware that many homeowners are armed and can react unpredictably when someone invades their home. Finally, since he expects a relatively small payoff for any particular burglary, he will tend to be more easily dissuaded.

While we would not attempt to generalize so easily on the behavior of the rapist or deviate who breaks in intending to commit a violent act, we still believe the effect of an alarm sounding throughout the neighborhood would work in your favor.

Of course, a home alarm system can be extended to include central station monitoring and response, usually in addition to the local alarm. It should be understood this additional service is always available when the dangers or values warrant it. The digital dialer is commonly used today and could cost as little as $15 additional per month. For the homeowner who has highly valued goods to protect, or feels a special concern for personal safety, this is a worthwhile improvement, but one which adds greater complexity to the management of the system and response to alarms. We will have more to say about this later in the chapter.

A note on central stations: For practical reasons, virtually no central station alarm companies will agree to monitor or service an alarm system that it did not install or approve in advance with the installing company. The only exception would be if the system was installed by another central station alarm company to industry standards.

One of the first questions a homeowner confronts in making decisions on alarm systems is whether to purchase a 'packaged alarm' and install it himself or have the job done professionally. In response to today's growing concern for security, the consumer electronics industry is bursting with alarm products for the home. There is a bewildering array of do-it-yourself, self-contained alarm installations designed to detect fire or intrusion and ring a local bell.

Some of this equipment is excellent. But beware. Some is truly worthless; worse, even dangerous. Once installed, it tends to be relied upon when, in fact, it will fail. Some equipment is so poorly designed that you could install a fire alarm system that would actually start the fire it was designed to detect.

Do-it-yourself...

Even the best 'package' alarm will give inadequate protection if its operation is poorly understood by the person who buys it. One advantage of working

...can be done, but...

...you'd better be good.

with a professional alarm company is the education about alarms that usually rubs off in the process.

The requirements of a reliable home alarm, even one that will only ring bells and turn on lights, are as technically demanding as those of the average business. Indeed, the home presents several special problems which make it in some ways more difficult than the business property to properly fit with alarms.

Alarm proposals are usually free.

We think it is especially worthwhile to consult a professional alarm company before you buy anything. Usually there is no cost for this consultation.

Take advantage!

If you decide to seek professional help, you can probably find a local alarm company to design and install a home alarm system for you under various purchase formats. You may buy it outright and have the company service the system as needed, billing you only for service calls. Or you can choose to use the commercial alarm format, paying an installation fee at the beginning and a monthly bill to pay for service and amortization of the equipment.

Often such a system will cost only a small amount more than the equivalent equipment you can buy in an electronics store. You will know that the equipment will be serviced by those who installed it. The experienced alarm installer will also know of many useful home alarm features beyond your knowledge which can be added at relatively little cost.

In cooperation with a local alarm company, you can pick out the level of protection that fits your budget. A modest system properly installed by a professional company can be expanded in future years as your budget allows. Here we will describe elements of a full home security system and place before you a view of some options.

Basic Elements of the Full Home Alarm System

The home alarm system tends to be oriented toward a more flexible, complex pattern of living in which the responsibility for operating the system falls collectively on a circle of people often including two or more adults, their children and sometimes friends and relatives, all of whom can be expected to have other things on their minds than alarms.

Living patterns are hard to change...

With a commercial alarm, protection is usually geared to an unoccupied premise at night. In the home the alarm must be adapted to provide various levels of protection with some protection active at all times. The protection covers both burglary and fire and must be adaptable to many conditions.

We are therefore going to look at several functions, some of which are found in commercial alarms and some of which are not:

...make the alarm adapt.

- an active, 24-hour, fire protection system.

- a high level of identification that the premises are protected against burglary.

- a burglar alarm response system to ward off intruders, alert occupants of an alarm condition and attract attention in the neighborhood.

- a home control center for activating and deactivating the different elements of the system with multiple on/off locations.

- a dual perimeter/interior system of intrusion detection making the system adaptable to provide protection while the home is vacant or occupied.

- panic buttons and other special devices designed specifically to promote the safety of occupants.

In most ways these elements are analogous to the equipment and concepts used in commercial alarms. A brief explanation follows:

24-Hour Fire Protection

Fire alarm systems for the home are not usually designed for approval by rating agencies. Instead, the standard of installation is guided by practical service and cost-efficiency.

Because of the need for early warning (most fire fatalities are caused by smoke inhalation as people sleep), **smoke detection** is usually used in the home as the primary form of fire protection. There should be an audible smoke detector in each sleeping area.

We discussed these devices thoroughly in the previous chapter. Regarding the use of smoke detectors in the home, these are some points to observe:

Testing is your responsibility...

- To minimize the number of **smoke detectors,** they are usually located in central areas such as hallways and attics.

- Location and sensitivity levels of smoke detectors have to allow for smoke from kitchens, fireplaces and cigarettes so they will not be continually set off by daily household occurrences.

- Testing by introducing smoke into smoke detectors should be done on a regular basis.

...Make it a part of house work.

- **Heat detectors** can be added in special areas where smoke detectors are unworkable or fire is likely to break out— around hot water heaters, in kitchens, near electrical junction boxes, etc.

- **Alarm bells** on fire systems are configured to ring differently than burglary systems connected to the same bells (steady ringing for burglary and pulsed for fire is common). Neighbors should be made aware of this distinction.

- With the exception of the bells, all fire alarm **wiring** should be completely separate from any other electrical system including the burglary system.

- Standby power is considered essential since fires often break out in the electrical system.

- Fire alarm circuits should be 'active' so that any wire that burns through or any failure will alarm the system.

- When adding iron bars to your house or sealing windows for burglary protection, be sure to leave fire **escape routes** that are well known to your family. A simple fire drill can save a life in the confusion of an actual emergency.

Alarm Identification

It is generally accepted as good procedure to prominently display signs or decals bearing the logo of the alarm company at every entrance to the house.

Burglars have fears too!

In the past, some homeowners shied away from this practice on the theory that it only announced to burglars that the house contained valuable goods.

If this was ever true, it can no longer be considered so. Alarm systems are so prevalent today that the conclusion most likely to be drawn from the warning sign is that the homeowner has a concern for protection.

To an outside observer, most homes in a given neighborhood appear to contain about the same value in personal goods. The knowledge that he will be contending with a burglar alarm system will sometimes deter a burglar who can expect to find the same items in a nearby home without the additional hazard.

The Home Alarm Center—Brain of the System

Instead of the alarm transmitting device found in commercial systems, home alarms usually have a panel from which normal alarm functions are controlled.

The primary purpose of the home alarm center is to MAKE THE ALARM SIMPLE and effective for the daily use of a family, reducing the incidence of false alarms caused through confusion.

Home Control Panel

In a basic system, the home alarm center is simply an integrated circuit built into the alarm bell housing and controlled by key stations.

In a more complex system the control center has its own housing usually located in the master bedroom or near the main entry in a closet or as a flush-mounted panel. These units:

- Provide standby power for the alarm systems.

- Provide indicator lights to tell the status of all the systems (on/off, clear-to-set, power and status of special features and fire alarm).

Digital dialers have home panels built-in.

Digital Keypad

As the alarm system grows in complexity, so does the home alarm center. Instead of a simple on/off switch, alarms can be controlled by multiple key stations or digital 'keypad' controls (with or without time delays) that can arm or disarm the alarm system or remove specific devices from the active protection circuit. The variations are almost unlimited. However, the only rule is, 'simplicity always works best.' Some practical considerations include:

- With time-delay units, you have a preset amount of time after the alarm is set (30 to 60 seconds) before it arms itself. You have that much time to exit the premises. Likewise, you have a preset period of time upon entering to disarm the alarm.

- If keys or digital controls are used, you set the alarm from outside. The round-shank standard alarm key is best as it is harder to defeat.

- All external controls and bell housings to alarms should have **tamper switches** to sound the alarm if disturbed.

A word of caution: Home alarm systems are being designed to give conveniences that have nothing to do with security. This is fine as long as it does not interfere with the efficient operation of the security system. If you want these other conveniences you are probably better off to get a second unit that is distinct from the security system.

Interior/Perimeter Protection Circuits

Perimeter protection consists of a ring of protection devices on the perimeter of your house to warn you of an attempt at entry and ward off that attempt before it has been accomplished. It consists of:

For maximum flexibility...

- Contacts on exterior doors.

...the perimeter alarm is the main system.

Aluminum Screen

- Alarm screens on openable windows (and some fixed windows vulnerable to intrusion). Where alarm screens prove to be too expensive or unscreened plate glass or decorative windows are prevalent, apply foil when aesthetics permit, or shock sensors when looks present a consideration.

- Alarm screens on doors with openable glass inserts (common kitchen door).

- Contacts on small openable windows (windows that need to be opened for ventilation are double contacted to allow two settings).

- Contact switches on outside attic hatches and outside attached tool rooms or garages.

The interior devices add secondary protection when desired.

Because full perimeter protection is usually impractical, a second alarm circuit, **interior protection**, is connected to devices installed within the house to detect a breach of the perimeter.

Interior alarm systems include:

Under-Carpet Switch

- Separate on/off control so the perimeter system can be left on when the house is occupied and people may be moving through the area normally covered by the interior protection system.

- Floor switches installed underneath carpet or on independent mats. These are usually placed near points of entry such as a vulnerable window, or in a central traffic area. (Good ones are quite reliable, but beware of the placement of bulky models which can show a bulge through carpet.)

In-Wall 'Laser'

- Recessed photoelectric cells or lasers (that have the appearance of an electrical wall socket) as well as self- contained infrared or ultrasonic units to cover specific rooms or high-traffic areas.

- Specially designed microwave units installed in attics and pointed downward will penetrate the ceiling to cover several rooms at once. This is generally considered inadvisable, however.

- Interior 'hidden' switches recessed into door jambs can be used on specific interior doors or cabinets where valuables are kept.

Hidden Magnetic Switch

- Interior on/off control in the master bedroom leaving clear access to a bathroom. (This allows the full use of the security system after you retire at night).

Note: With all internal devices house pets must be taken into consideration. Their movements can set off switches under carpet, motion detectors, infrared detectors and photoelectric detectors. All these things are dependent on the height and weight of the pet. You need to consult with the alarm company.

Panic Buttons and Other Special Devices

Panic buttons are usually installed in the master bedroom and at least one other location central to the house. These buttons are never off and activate the burglar alarm any time they are pushed.

Other special applications frequently found valuable in the home are:

Self-Contained Ultrasonic Unit

- Specially designed devices installed in **swimming pools** to alarm when a disturbance occurs such as a child or animal falling in. If connected to your alarm system these should have their own specific sounding device to differentiate from other alarms.

Gates: usually more hassle than they're worth

- **Outside gate contacts**, consisting of heavy-duty, weathertight switches to avoid false alarms. (Any gate at the outside of your property must have a separate control switch to allow entry and exit without disturbing the house alarm.)

Medical Alert Receiver

A note on medical alert: A medical alert system consists of a remote button carried on the person which, when pressed, sends a signal to a receiver on the premises, usually within several hundred feet. The receiver activates an alarm transmitter, usually a digital dialer which relays the signal to the intended receiving station.

A medical alert system is not a part of a burglary system. It is always connected independently off-premises to some form of central receiving station.

Of course, if you have an alarm this could be the same company that provides your alarm service.

The alarm signal is handled as a medical emergency by the central station and an ambulance is dispatched. Many central stations do not handle medical emergency alarms due to the liability involved.

Alarm Response

What happens when an alarm goes off? In some ways, this is both the easiest and most difficult question concerning home alarms. It is easy because a few simple guidelines which we are about to discuss can make alarm handling quite efficient. Yet, as we said earlier, this is a question of personal safety. And this makes it difficult. No two people react the same in the face of danger.

Panic buttons give peace of mind.

The answer to this question is so personal we would like to sidestep it completely. Yet we suspect a solid, convincing answer would be worth more than a thousand words of technical know-how.

We can forewarn you with certainty that your alarm will go off and that most of the time when it does there will not be anyone there. If you thought installing an alarm system was going to eliminate your having to confront the potential of an intruder in your home, you are wrong. While it may, in fact, provide more protection from an intruder, you are going to have to confront the possibility of one every time it false alarms. These practice confrontations with false alarms may make you better equipped to deal with the real thing. So what do you do when it happens?

To keep fear in its place ...

Here are several rules of thumb:

- Treat every alarm as if it were real until proven otherwise.

- First determine if it is a fire or burglar alarm.

- If it is fire, and you are connected to the central station, immediately investigate the situation, being aware that the alarm company is probably dispatching the fire department to your house while you are doing so. When you have determined you are in no immediate danger, get the alarm company on the phone to save the fire department an unnecessary emergency response that could endanger others. If you are not connected to the central station call the fire department directly. Always have emergency procedures clear to you and your family.

- If it is burglary, and you are connected to the central station, they will probably be calling you. If you are not connected to the central station, call the police and ask them what to do (do you have the number handy and a phone near your bed?).

Plan ahead ...

In all events, do not overreact. It could be a member of your family. It could be a malfunction in the system. It could be an actual intrusion. In any of those

And ...

circumstances, allowing the alarm system to continue to ring for a few minutes is not going to do any harm and could very well discourage an actual intruder. We can't tell you when it is safe to investigate your house. A common practice is to turn on the bedroom lights, allow the alarm system to continue to ring for a time, get the reassuring support of your alarm company or police station on the phone and deal with the situation as you and they determine is practical. If neither the police nor the alarm company is appropriate, getting a friend on the phone can also be reassuring. No friend minds being awakened if you are in fear.

Putting emotional issues aside, there are a number of technical considerations that must receive special treatment to facilitate response to alarms in the home:

**Follow sound
practices** • • • • •

- Have your alarm system designed so that you have some control over it from the master bedroom area. On the more elaborate end, it is possible to design lights into your home alarm panel that will tell you where in your house and by what type of device an alarm was tripped. This can help you in determining how to react. It is also possible to turn on various lights from the home panel.

- When an alarm is connected to a central station, a **time delay** is often built in to hold the silent transmitter from sending a signal to the central station for 30 to 60 seconds after the alarm on the premises has been activated. This is to allow you to control false alarms before the central station and the police are alerted, necessitating lengthy explanations.

- **Bells, horns, buzzers, beepers** and **sirens** should be installed to be audible from both inside and outside the house. Multiple devices can be used if necessary.

- Sounding devices should be secured to a solid wall with wiring protected from tampering. Any alarm wiring on the outside of the house should be in conduit.

- All alarm bells attached to the outside should be in tamper- proof alarm housings designed to take several minutes to open and, when armed, alarm in the initial stages of unscrewing the housing. One of the best is the UL Approved bell box.

- Usually a five- to fifteen-minute reset/cutoff function is installed to reset the alarm system after an alarm if no one does this manually. If the system will not reset, the sounding device will be shut off after the preset period of time.

- The sounding device must be equipped to emit a distinctive sound for each alarm circuit with a different purpose (fire, burglary, pool safety). Occupants should be well trained to recognize each sound and have a plan of action for each.

- In addition to ringing bells, alarm systems can easily be equipped to turn on outside lights. This can be an added deterrent that does not increase the annoyance burden on neighbors.

- A new development in home alarms is the use of **voice simulators**, pre-programmed microchips with the ability to broadcast a message in spoken words through a speaker. At its present level of sophistication, the voice simulator can produce a fairly complex sentence to tell you of any condition that would otherwise be indicated by a warning light or alarm. This can be a reminder, such as those now being used in automobiles to tell you, 'Your lights are on,' or a response to an alarm with the voice programmed to say, 'Intruder,' 'Fire,' or even to simulate the sound of a barking dog. The flexibility and practicality of these electronic voice simulators will assuredly increase rapidly. Voice simulators can be looked for in a wide variety of formats in the near future.

Installation of Home Alarms

When planning an electronic alarm installation in a home, the first consideration is the placement of wiring from the system's trip-points to the alarm transmitter. Instead of running wires along the inside walls of your house, as is done in the typical business, the alarm installers will have to use the crawl spaces in the attic and beneath flooring, drilling access holes with long bits to window sills and door jambs.

Because this makes the job considerably more demanding than in a business, extra time spent in the design and engineering will usually pay off in a savings of installation time. If it seems to you that the installation charge proposed by the alarm company is too high, you can ask to be charged for actual hours worked usually at a rate of $25 to $45 per hour (this rate covers the additional costs of engineering work done at the alarm company plus overhead).

It is advisable to go over the installation plan carefully before the work begins so you will know in advance how much disruption the job is likely to entail, where problems may occur and what pieces of alarm equipment cannot be hidden. Sometimes you may have to lower your expectations a bit.

Pre-Wiring During Construction

There is no better time to install alarm wiring than when a house is being constructed. The wiring can be easily hidden, junction boxes and service points can be placed exactly where most useful and the whole job will cost considerably less.

Whether you are an architect, a contractor or owner, consult a central station alarm company experienced in servicing home alarms in the planning stage. Some alarm companies will provide this service at no cost, in hopes of getting your business. All you need to do is provide the architectural drawings.

You can only pre-wire a house once.

In a tremendous number of pre-wired homes, the wiring is not done properly, resulting in the loss of all the advantages gained by pre-wiring. Often the alarm company called in to install the system finds that little or nothing of the pre-wiring is usable, necessitating costly reinstallation. This problem can be avoided by working during construction with an alarm company that maintains its own qualified engineering department.

Wireless Alarm Systems

Where pre-wiring is not available the architecture of a home often makes wiring an alarm system extremely difficult, expensive and impossible to hide. As a solution to this problem, wireless alarm systems have been developed.

They fall into two main categories.

One is the use of the existing 110-volt electrical wiring to carry signals from one area of the house. to another. This system can be effective. But, because your electrical system is not truly separated from neighboring houses sharing the same electrical transformer, false-alarm problems could arise should a similar system be used nearby. Evaluate carefully before buying.

The other type of wireless alarm uses small radio transmitters located at each alarm device trip-point (window or door switch, motion detector, etc.). When activated, it transmits a signal to a built-in receiving transmitter in the home panel. The home panel transmits the alarm just as it would in a wired system. This type of alarm system can also be effective and help in simplifying wiring problems. Its weakness is that the signal transmitters in the devices are usually battery powered. Failure to carefully monitor the condition of these batteries (which may be numerous) can lead to a failure or ineffectiveness of your alarm system. It often happens that such an alarm system is so difficult to maintain it is allowed to become defunct.

Some wireless devices can be incorporated into an otherwise wired alarm system with great success when the device is maintained by a low-voltage transformer, ensuring a constant charged-battery condition. This eliminates the need for battery replacement.

What Do Home Alarms Cost?

$500 and up to install.

Considering the amount of square-footage protected, home alarm installations usually require more equipment than business alarms. This is partly because home alarm systems include devices such as smoke detectors, inside-outside controls and panic alarms often not found in business alarms. Also, homes usually have more windows and doors to protect than businesses of similar size.

Once installed, home alarms, whether connected to a central station or not, are charged a monthly service rate based on the number of devices and extent of the installation. This could be $25 to $50 per month if the system is not connected to a central station and $12 to $15 more if connected to a central station using a digital dialer, which is quite popular in home alarms.

Easily $1,000 for a good one.

These figures will vary depending on the quality of the alarm company, the equipment used and the central station involved, as well as the amount charged for installation. Be sure you read any alarm agreement for length of term.

The monthly service charge, in addition to amortizing costs not billed for in the installation charge, is based upon an anticipated service burden for the system.

What is Minimum Security in the Home?

The lowest-cost approach which will give you a significant amount of protection, is the use of independent, self-contained devices, both for burglary and fire.

Battery powered smoke detectors are available in most hardware stores for $15 and up. They are quite effective, can be installed in a few minutes and will operate up to a year on one battery. The better ones give a warning when the battery is low.

Many cities require dwellings to have wired smoke detectors based on experience showing that most people neglect to replace the batteries when they fail.

Self-contained burglary devices are a little more complex. All such units consist of a power supply (with or without battery backup), a detection device and a bell or horn sometimes backed up with lights. Self-contained devices detect intrusion by ultrasonic, infrared, microwave and microphones tuned to the specific sounds of break-ins. These devices are of varying quality depending on the engineering of the electronics. The cost for those units generally accepted to be reliable varies from several hundred to a thousand dollars. Most of them can be connected to external bells, controls or auxiliary alarm devices.

The value of self-contained systems is in their versatility, portability and simplicity of installation. Their value is limited, however, because: the home is usually penetrated before they go off; they are often easily put out of service by a burglar, and, unless connected to auxiliary devices, they will sound an alarm only inside your house.

Unless you know that you have experience with electronics and alarms, we do not recommend the do-it-yourself approach to home security.

It is probably a wiser course, before you spend any money on intrusion alarm devices, to investigate the services that can be offered by a professional alarm company.

Remember, in addition to proper design and installation, alarms need service from time to time. Many expensive alarms sit unused because something is wrong and reliable servicing has not been provided for.

Whatever your price range, achieving good security in the home requires a balance between good design, quality equipment, proper installation, practical operation and provision for prompt, reliable servicing.

Part II
LIVING WITH ALARMS

BUYING THE ALARM

We will begin this section with a simple checklist of direct steps to make the purchase of the right alarm system as painless as possible. This will be followed by discussion of some problem areas in the selection and negotiation process.

Checklist of Practical Steps

☐ Call three alarm companies to submit bids.

☐ Provide an 8½ by 11 inch rough floorplan or blueprint of your building.

☐ Define your hours of operation including variances for holiday schedules.

☐ Identify your responsible personnel.

☐ Consult your insurance agent (he is your partner at risk).

☐ Have copies of the alarm proposal sent to your insurance agent. Make sure your insurance carrier approves the plan you select (see 'The Most Costly Mistake').

☐ Discuss the differences in proposals with your insurance agent.

☐ Discuss the differences in proposals with the alarm companies at a management level where you are confident you are receiving clear information.

☐ Draw your best conclusions of what you want from each proposal with any new ideas you have found and ask for a revised bid from each company.

☐ Locate the central station address of each alarm company to see how close it is to you.

☐ Make two lists of the alarm companies, one in the order you think they can really provide the services to you they offer and the other in the order of cost.

☐ Evaluate the central station operation of the company you are interested in.

☐ Ask the one you feel can best provide services to justify its cost to you, whether or not it is the most expensive.

☐ Decide which bid provides the protection you want at the least cost (see 'The Low-Bid Problem').

The contract:

Then negotiate on the basis that the cost offered is specific to the elements in your alarm but general within certain limits. These limits depend on what kind of a customer you offer. An alarm system that is a nightmare to install and a burden to service is usually a loser to an alarm company at any reasonable price. Simple, clean jobs installed properly are the lifeblood of an alarm company. You fall somewhere in between the two. That range is where you can negotiate. The financial difference is usually from none to 20% in either the installation or the monthly rate, not usually both.

Negotiating is OK

Once you have contracted for a system, it is always a good idea to be present when the installation supervisor comes out to go over the job. Find out how he feels about what his alarm salesperson has sold you. He may have a better idea. Determine with him the exact location and appearance of alarm devices and wiring to avoid conflicts later during installation. Don't be upset about making changes at this point. This is a normal part of the complex process of designing and installing the right security system.

Flexibility after signing is OK too

The Most Costly Mistake

This is a three-party decision

The most costly mistake you can make is to sign a contract for alarm service and have the system installed prior to receiving written agreement from your insurance agency that the service will qualify for insurance. The insurance underwriter can reject the whole job even if there is nothing wrong with it. You must then start over at the beginning, buying a completely new alarm system (down to the wire and screws) and negotiating your way out of the five-year agreement you just entered.

A note on insurance: Insurance agents do not underwrite insurance. Underwriters do, and they often have a very different idea than their agents of what is acceptable for insurance.

A common stumbling block is the discrepancy between what you really need for protection, what the alarm company wants to sell you and what the insurance underwriter wants you to install.

The major difference will arise when the insurance company is asking for UL Certification without really understanding whether the difference in security will justify the added costs. Jumping from one level of certification to another can increase the monthly service charge up to $50.

It will probably be up to you to talk to the alarm company manager to ascertain the real differences involved in certification of your alarm and then to try to form a compromise with the insurance underwriter as well as the agent. The usual approach is to:

A. Inform the insurance company that while the alarm is not UL Certified, it is installed to UL specifications for the certification they are asking for and is connected to a UL Approved alarm company.

B. If this does not satisfy the underwriter, talk to the alarm company and try to work out a lower certification that will provide what the insurance company really needs.

C. If that fails, you can shop the insurance. But you must understand that insurance is underwritten by companies far removed from your business and they must work out standards based on average losses. To have the insurance you need, you may have to live with a little higher alarm service rate than you would like.

The Low-Bid Problem

In a competitive world, alarm company salesmen will often leave important items off a proposed contract because of the fear of overbidding competitors.

Before selecting any bid make sure it includes:

- All phone line charges. Often the phone company will be asked to bill you directly. This can add significantly to your cost.

- Only UL Certified equipment and installation techniques. The added cost is easily justified if you ever decide to ask for certification.

- Adequate zoning of alarm circuits.

- Self-testing, clearing and setting features in the alarm system.

If you find that alarm bids differ greatly in the type of equipment being proposed, this is not always a result of different engineering approaches, but can be a cost-based decision designed to keep the bidding company competitive. While it is wrong to sacrifice engineering for cost, it is nevertheless commonly done. Particularly if you see different pieces of equipment being recommended for the same application, you may want to refer to sections of this book on those devices to see if there is any reason one may be better than the other.

Taking Over an Old Alarm

A dangerous gamble

If you have moved into a building where a previous company had an alarm installed you can sometimes use the existing wiring to save yourself some installation costs. However, if you are changing alarm companies, the old company may take a dim view of you letting their competitor attach equipment to their wiring (remember, standard alarm agreements are for service and the equipment belongs to the alarm company).

Consider only if the system is new

Upon finding that you cannot be persuaded to use their services, the old company may send a technician to remove its equipment and disable any wiring being left behind. You can usually avoid this by asking the new alarm company to handle the situation. They will usually remove the major equipment and leave it at your front door for the other company to pick up, thereby avoiding the possibility of anyone damaging the wiring they intend to use.

In some cases, your new alarm company may have permission to use the old equipment and wiring, but may choose not to. Generally, you should accept this judgment. Alarm companies do not mind saving labor and will use existing equipment when it is usable. But older wiring is often the source of more trouble than it is worth. Years of service tinkering can lead to a false alarm problem.

Conversely, if you are leaving a building with an unexpired alarm agreement, sometimes the only way out of paying the remainder of the agreement is to have the new tenant take over the alarm system agreement.

Understanding the Alarm Contract

Not a basic consumer contract ...

The length of your agreement must be negotiated on the basis of the estimated time at the location. If you think you are going to be at the location less than five years, which is the term of the basic agreement, work out an agreement that allows you the five-year rate with any differences as a balloon payment at the end. Place the difference in an escrow account if you have to to guarantee it. But stipulate these conditions: A) You may continue the alarm if you stay longer than expected and when the five years is reached, the money is yours (it can be paid back in payments as the five-year date approaches); B) if a new tenant takes over your alarm at the same location, you are relieved of the responsibility for a balloon payment; C) if you contract for alarm service at a new location with the same company, you will pay a reduced amount of the balloon payment (50%).

Find out:

Although this may seem stiff, keep in mind that if the alarm company must remove the equipment from your building in two years and abandon the system, it will make little or no profit.

Be sure phone line charges are included in the costs.

Ask that the system be installed to UL specifications even if it is not going to be certified (put it on the agreement).

Before signing the agreement, clearly list every item of the alarm system and the service to be provided on the alarm agreement. It is important to have a complete understanding between you and the alarm company to create good security.

Service or lease?

If you are one who never reads the fine print of a contract, make an exception in this case because the contract most alarm companies use is a tough one, with severe limits on the alarm company's liability for your losses.

What do you own?

Alarm companies operate under tremendous potential liabilities, many of which are completely out of their control. Among these: phone lines which must be maintained by telephone companies, the actions of subscribers' employees, the compromises subscribers make to hold security costs down, leaving weaknesses in the security system.

What is their liability?

There are very few insurance companies in the world that will insure alarm companies against all liability. And yet, no alarm company can operate without insurance. Therefore, these insurance companies insist on clauses which set clear limits on the alarm company's liability.

The basic philosophy involved is that alarm companies do not provide insurance against loss, they provide a service to help you prevent it, and one that cannot be guaranteed. If you want insurance, you must buy insurance.

Sample Protection Plan With Costs

To give you an idea of the relative cost of a wide variety of alarm equipment, and to demonstrate how you can prepare a sample drawing of your own building when buying an alarm, we have created an imaginary, but quite typical, building and drawn a sample alarm proposal for it.

This sample is a 300' by 600' office and manufacturing facility containing merchandise and equipment of relatively high value. The burglary protection is broken into three separate alarm systems to accommodate the pattern of operation of such a facility. One system covers corporate offices, which operate on a normal daytime schedule. A second covers the warehouse, which can be closed to company personnel during night shift operations. A third covers the manufacturing area which usually opens earlier and operates until later in the evening.

An elaborate alarm system ...

Our example building is equipped with a water sprinkler system. It is protected by monitoring for water flow in the system and the On/Off condition of its valves. Smoke detectors were added in a computer room to give early warning.

But basic in approach

Because the alarm system is a complex one, it is priced as if installed by a large, full-line security company and is, therefore, moderately high in the individual component cost both in installation and monthly service, reflecting greater engineering and supervisory capabilities. Other factors affecting cost of the system would be long wire runs and a high standard of design integrity including such things as protective conduit on wiring.

Scale it to your needs.

Any of the systems alone installed in a smaller facility by a fairly small but competent alarm company could cost 25% less in both installation and monthly service charge. Also, the prices in our illustration would normally be negotiable perhaps up to 20% of the total cost.

Cost includes all telephone company installation and line charges with the exception of the digital dialer coupling, which is usually billed directly to the subscriber by the phone company. The standard proposal also includes all service, maintenance, alarm response, monthly testing and approval of fire alarms, and a monthly notification of all openings, closings and irregular entries. Prices are based on a five-year agreement.

To obtain Underwriters' Laboratories approval, the burglary systems would cost more depending on the grade and extent of the certification. This could cost considerably more per month even if no additional devices were needed.

All the zones of the systems are individually indicated both on the premises and back at the central station.

The Digital Dialer backup is designed to provide an alternate alarm line to the central station if the Multiplex Line should fail.

Rough Floorplan Showing Proposed Protection

The first of the two drawings on the following pages represents the bare floorplan of the building. The alarm designer would work with such a drawing to develop a protection plan. The second drawing is the final proposal, showing the location of each piece of equipment in the system. Each item on the drawing is coded for reference to the itemized list that follows.

When buying an alarm system it is a good practice to ask for a floorplan such as this and keep it as a permanent record for use later in testing and servicing of alarm equipment.

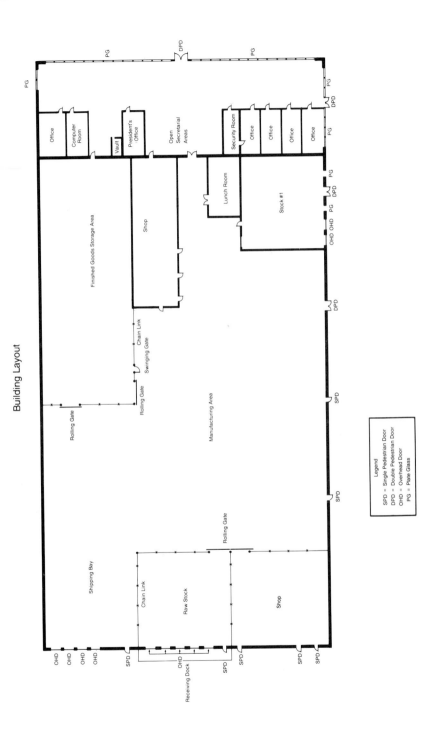

Building Layout

Legend

SPD = Single Pedestrian Door
DPD = Double Pedestrian Door
OHD = Overhead Door
PG = Plate Glass

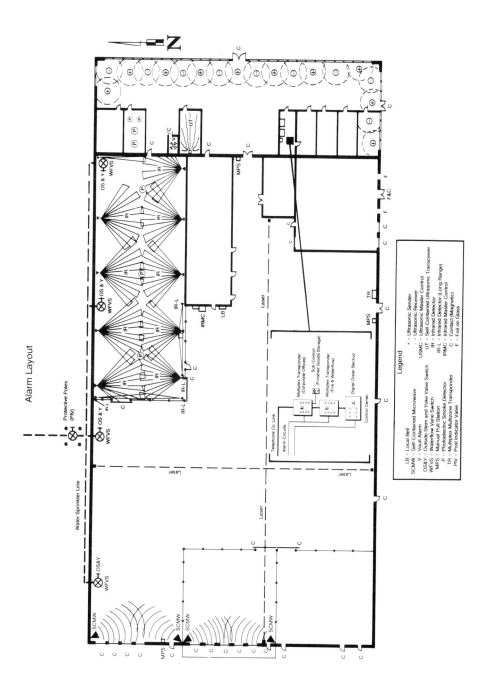

Alarm Layout

Proposed Protection Plan With Itemized Cost

Burglary Detection

Three Independent Supervised Systems:
Corporate Offices, Warehouse, Manufacturing Area
Installed to Meet or Exceed UL Specifications
But Not UL Certified

Quantity	Device and Location	Installation	Per Mo.
	ALL-SYSTEM TELEPHONE CONNECTION		
1	Multiplex Telephone Co. Connection, security room, (services all burglar and fire alarms)	350.00	45.00
1	Backup Digital Dialer on separate phone line, with 3 burglary zones, 2 fire zones and 1 valve tampering zone	180.00	10.00
	SYSTEM ONE—CORPORATE OFFICES — SUPERVISED		
	Supervised 7:30 a.m. opening to 5:30 p.m. closing		
1	Multiplex Multizone Transponder (TR), security room, with Day-Night Circuit Monitoring and Central Station Acknowledge, Power Indication, Battery Indication, End of Line Resistance	200.00	40.00
1	Local Bell (LB), manufacturing area	170.00	5.00
	ZONE ONE—TRANSPONDER ONE		
1	Multizone Ultrasonic Master Control (USMC), security room	200.00	35.00
19	Ultrasonic Transducers (+–) throughout @ $50	950.00	38.00
	ZONE TWO—TRANSPONDER ONE		
1	Ultrasonic Transceiver/Self-Contained (UT), president's office	90.00	3.00

	ZONE THREE—TRANSPONDER ONE		
2	Contacts, double pedestrian door, front	75.00	1.00
2	Contacts, double pedestrian doors, south side	75.00	1.00
	ZONE FOUR—TRANSPONDER ONE		
2	Contacts, double pedestrian door, manufacturing	75.00	1.00
1	Contact, single pedestrian door, manufacturing	50.00	.50
1	Contact, single pedestrian door, warehouse	50.00	.50
	ZONE FIVE—TRANSPONDER ONE		
1	Vault Alarm (V), vault	150.00	20.00
1	Vault Door, contact, vault	100.00	3.00
		$2,185.00	**$148.00**

Quantity	Device and Location	Installation	Per Mo.

SYSTEM TWO—WAREHOUSE—FINISHED GOODS
Supervised 7:30 a.m. to 5:30 p.m.

1	Multiplex Substation Control warehouse, with Day-Night, Acknowledge, Power Indication, end of line resistor	$125.00	$15.00
	ZONES SIX & SEVEN—TRANSPONDER ONE		
1	Multizone Infrared Master Control (IRMC)	150.00	20.00
14	Infrared Detectors (IR), throughout warehouse @ $75	1050.00	56.00
	ZONE EIGHT—TRANSPONDER ONE		
1	Contact, single pedestrian door, offices	50.00	.50
1	Contact, single pedestrian gate, manufacturing	70.00	1.00
2	Contacts, gates, manufacturing	140.00	2.00
	Wire and Connectors as needed (all wiring below 12' to be in conduit)	300.00	
		$ 1885.00	**94.50**

Quantity	Device and Location	Installation	Per Mo.
	SYSTEM THREE—MANUFACTURING **Supervised 6:30 a.m. to 12 p.m.**		
1	Multiplex Multizone Transponder, employee exit, Day-Night, Acknowledge, Power and Battery Indication, end of line resistor	$ 200.00	$ 40.00
	ZONE ONE		
2	Microwave Units/Self Contained (SCMW), shipping	250.00	16.00
	ZONE TWO		
2	Microwave Units/Self Contained, raw stock	250.00	16.00
	ZONE THREE		
2	Photoelectric Cells, 'Lasers,' manufacturing	250.00	12.00
	ZONE FOUR		
2	Contacts, double pedestrian door, employee entrance	75.00	1.00
2	Contacts and Foil, double pedestrian door, stock #1	95.00	2.00
2	Glass plates, foil, stock #1	50.00	1.00
2	Contacts, overhead doors, stock #1	150.00	2.00
2	Contacts, single pedestrian doors, manufacturing south side	150.00	2.00
	ZONE FIVE		
5	Contacts, single pedestrian doors, rear	250.00	2.50
9	Contacts, overhead doors, rear	675.00	9.00
	ZONE SIX		
1	Contact, single pedestrian door, stock #1	50.00	.50
1	Contact, loading door, stock #1	75.00	1.00
1	Contact, gate, raw stock	70.00	1.00
1	Contact, gate, shop	70.00	1.00
	Wire and connectors as needed (all wire below 12' in conduit)	450.00	
		$3,110.00	**$107.00**

Fire Detection
24-Hour Fully Approved

Quantity	Device and Location	Installation	Per Mo.
1	Multiplex Multizone Transponder, security room (for all fire alarm systems)	$ 200.00	$ 40.00

SPRINKLER VALVE MONITORING SYSTEM

ZONE ONE

1	Outside Stem & Yoke Switch (OS&Y), pit (at city connection)	100.00	3.00
1	Post Indicator Valve Switch (PIV), north side — outside	75.00	3.00

ZONE TWO

4	Outside Stem & Yoke Switches, risers north side	400.00	12.00

WATER FLOW MONITORING SYSTEM

ZONE THREE

2	Water Flow Vane Switches (WFVS), in risers-west	$ 250.00	$ 12.00

ZONE FOUR

2	Water Flow Vane Switches, in risers-east	250.00	12.00

AUTOMATIC FIRE DETECTION SYSTEM

ZONE FIVE

2	Photoelectric Detectors (P), computer room, subfloor	$ 150.00	$ 10.00

ZONE SIX

1	Photoelectric Detector, computer room	75.00	5.00
1	Photoelectric Detector, computer room, above ceiling	75.00	5.00

ZONE SEVEN

3	Manual Pull Stations (MPS), manufacturing	350.00	12.00
3	Photoelectric Detectors, finished goods storage area.	225.00	15.00
	50 feet of trenching across blacktop to pit valve	1760.00	
	Wire and connectors as needed (@ $11 per foot)	550.00	

$ **4,460.00 129.00**

TOTALS

Total Cost, Burglary	7,180.00	349.50
Total Cost, Fire	4,460.00	129.00
Telephone Connection (Multiplex)	350.00	45.00
Telephone Backup (Digital Dialer)	180.00	10.00
Combined Total	**12,170.00**	**533.50**

Contracted Patrol Service to respond to all alarms and three exterior checks per night from 1 a.m. to 4 a.m.   150.00

Total Installation and Monthly Cost $ **12,170.00 683.50**

After the Agreement

When you have successfully concluded an alarm agreement, you will be asked to provide the alarm company with a complete schedule of your openings and closings for each separate system (including variations for holidays). This is not necessarily your daily business hours, but the actual time that the first employee arrives and the last employee departs.

In addition, you will have to determine which of your employees will be authorized to make decisions affecting the alarm and declare their names on a form such as the following:

AUTHORIZED PERSONNEL LIST

The following named persons are those authorized to make changes in the alarm system operation or authorized personnel.

Name	Home Address	Phone Number	Signature	Indentification Code
_____	_____	_____	_____	_____
_____	_____	_____	_____	_____

(Usually officers of the company who are legally empowered to act as company representatives.)

The above signatures are the only signatures the alarm company will recognize as empowered to make changes in its agreement with you.

The following named persons are authorized to have access to the premises during closed hours:

Name	Home Address	Phone Number	Signature	Indentification Code
_____	_____	_____	_____	_____
_____	_____	_____	_____	_____

(As few as possible, not necessarily those authorized to open or close.)

The following named persons are to be notified in the order listed in the event of an alarm:

Name	Home Address	Phone Number	Signature	Identification Code
_____	_____	_____	_____	_____
_____	_____	_____	_____	_____

(Nearby persons who are familiar with the alarm system and your company operation, with access keys and able to act as a company representative. These persons may be exposed to danger.)

All phone numbers must be kept updated and any terminated personnel must be immediately removed from the list of authorized personnel.

All changes in business operation or personnel must be in writing with signatures followed by the printed name with the exception of deletion of authorized personnel or late closings, which can be done by telephone by any of the personnel authorized.

WORKING WITH ALARMS

Day in and day out, the attention you pay to your alarm system can become the single greatest factor in its successful, trouble free performance.

In this section we are going to discuss the things which can only be done by you to ensure that you have the best possible security service. We hope to make you aware of some responsibilities which your insurance agent and alarm company may have neglected to tell you. We'll give you hints on how to set up procedures that will make the day-in, day-out job of living with an alarm as easy as possible. We'll also discuss negative procedures which compromise the security provided by your alarm system.

Giving your alarm system the attention it requires shouldn't be so taxing that you begin to wonder if it was all worth the trouble. Take the time in the beginning to set up the right procedures and to make sure all your employees understand and respect them. The rest will soon be habit.

At the outset, you should create a security file folder with a subfile 'Alarms' and include in it an outline of all alarm procedures for use by the person responsible for dealing with the alarm company. If your records are on computer, put this information there. This should be a 'sensitive' file with restricted access.

On a day-to-day basis, you should be conscious of the details involved in the following responsibilities:

Your responsibilities include • • • • •

- Clearing, testing and setting the alarm properly.

- Keeping the alarm company immediately updated of all personnel changes affecting security.

- Keeping the alarm company notified in advance, in writing if possible, of all schedule changes.

- Creating workable procedures for early openings, late closings, and entries at unusual hours so that these don't confuse the alarm company.

- Protecting alarm equipment from damage through carelessness by employees.

- Planning in advance what to do when there is an alarm and providing that information to the alarm company.

- Periodic testing of alarm components for range and sensitivity.

Turning the Alarm On/Off

You are responsible for turning your alarm system on or off at the proper time, notifying your alarm company of any change in this pattern and making regular tests to be sure the system is working as it should.

No alarm system—not even the most electronically sophisticated wonder that money can buy—will live up to its potential if treated carelessly during the simple act of opening and closing your business.

Yet, all too often, that is exactly what happens.

Changes of habit confuse ...

A common breach of security occurs when the last person out at night makes a simple mistake. Suppose you turn on your alarm system, check that it is working, open the door to leave, but then notice you left your brief case on your desk. You close the door, walk back to your desk and then rush out. The second time you open the door, an unexpected alarm shows up at the central station. The operator immediately calls your office, but nobody is there to answer. He may guess what happened, but, unable to reach you at home, he calls in the police. This could be the beginning of a bad relationship.

Can the central station interpret your signals?

Develop a list of procedures for the hours of entry and exit for the employees (the fewer the better) who are responsible for turning on and off the alarm or need to have access after closing hours. Impress on all who may come in or out on their own that they will be causing an alarm signal at the central station and they must take steps to keep the alarm from being transmitted to police. Post a list of hours and procedures beside the alarm transmitter to help ensure the job is done right. The instructions should include mention of all skylights or windows (such as in bathrooms) that may be opened during the day but need to be closed after working hours prior to setting the alarm (also include such things as turning off coffee pots, air conditioner, etc.).

Clearing and Setting the Alarm

Perhaps the greatest virtue of an alarm system is that it requires you to take special care that you are leaving a tightly-closed building at night.

Take a careful look at your alarm transmitter as you leave the building each night to see that the system is in order. Any protected door ajar, or a window not closed, will show up on the transmitter indicating that the system is not clear to be turned on for the night.

If you ignore this warning, you will be sacrificing alarm protection for the night. If circumstances require you to do that, you should always call the alarm company to tell them that is what you are doing. If your system is supervised, the central station operators will be anticipating a closing signal from you.

Take it seriously ...

If the alarm transmitter will not clear, take the time to search for the problem. Do not make the common mistake of having the alarm company send out a busy serviceman simply to close a door or window that was left open. You do what you can to find the problem first, then call the alarm company.

...or you'll never get it right.

We should add here that a similar problem can occur if a door not quite latched or a window barely cracked open gives a clear signal as you leave. Later at night, a draft can pull on a door or window moving it just enough to set off the trip-switch. If this happens, the alarm company will probably pass along any false alarm charges to you because the problem was caused by you. Passing on the charge to whoever failed to secure the window or door may be a tonic for carelessness.

Sometimes, of course, it may be a malfunction of alarm equipment, rather than anything you or your employees have done, that keeps an alarm system from clearing. If you cannot easily locate the problem, the proper course is to call the alarm company for a service visit.

A note on setting the alarm: The better alarm trans-mitters in use today have a light on the transmitter to indicate when the system is on and secure. However, some key operated alarm systems lack this feature, causing a great temptation to merely switch the key and leave, without any check.

Don't expect your alarm company to keep a platoon of service men standing by for your call. Because alarm servicemen are hit with many service calls during the hours when most businesses are closing, you can avoid considerable inconvenience by testing the alarm system about an hour before you close. Again, if you notice a malfunction just as you are leaving and don't take the time to correct it, you are making a conscious decision to forego security for that night.

Early Opening/Late Closing

Seldom is the morning opening or evening closing strictly adhered to. There are some simple rules of thumb used by most alarm companies to accommodate normal variations.

Is the central station number posted?

Closing: up to an hour early or an hour late in closing the usual procedure is followed. More than an hour either way, the alarm company should be called. Call ahead of time to tell the company when you plan to close and then again as you leave to confirm that you are closing. This may sound redundant, but remember how things work in the central station. Sometimes an operator will see your late closing signal but may not have been on duty when you called. He may think your closing signal is a burglary. UL certified systems often have tighter procedures for opening and closing hours than non-UL systems. Be sure to find out the operating procedures for the central station you select.

Opening: if you are opening more than 10 minutes early, you must call the alarm company, but up to an hour late you can do as usual. An early signal poses a difficult decision for alarm company personnel. Early morning burglaries are not uncommon. The operator must hear from you to know what is going on. They appreciate your concern.

Responding to Alarms

It is equally important to decide ahead of time exactly what should be done when there is an alarm. Invariably, you will be called late at night in your home. The central station operator tells you an alarm has come in and the police have been called. You dress and start the 20-minute drive to meet the police there.

This is where life is in danger.

Five minutes later, a police patrol car pulls up in front of your business. The officers wait for 10 minutes, but see no sign of activity around the dark facade of your building. So they leave on another call. You arrive 10 minutes later, unlock the front door and come face-to-face with the burglar who was at work inside all the time the patrol car was waiting outside.

There are various techniques to circumvent this problem:

- Provide the alarm company with keys.

- Contract with a patrol service to provide response.

- Use the closest available employee of your company (this should be someone with real common sense, as danger could be involved).

- Notify the alarm company how long you expect to take to reach the location so that you can be met.

- Call the alarm company back from a nearby phone when you reach the location if no one is present. The police will usually return.

- Don't go in alone. Meet police or alarm patrol outside.

During the adjustment period of a new alarm system or when service problems are occurring, late-night phone calls can become such an aggravation that you may be tempted to tell the central station to ignore the alarm or to call out the police as you fade back into sleep and forget the problem. This approach amounts to a conscious decision to abandon security. If you don't take your alarm system seriously, neither will the police or the central station.

Updating Your Alarm Company

Terminate an employee?

You should make a practice of regularly notifying your alarm company of changes in the list of employees who are authorized to carry keys to your building and who may be communicating with the alarm company when opening or closing the premises.

Let them know.

Do not ask employees to communicate variations to the alarm company unless they are authorized in writing to do so. Alarm companies cannot accept spur-of-the-moment alterations in their agreement with you except by personnel authorized to do so and with UL Approved systems only in writing and in advance.

Testing Your Own Alarm System

No alarm system can be considered dependable unless it is tested daily and no one is in a better position to do so than you.

The only way to know if it works

By conducting a few simple tests periodically, you will gain a real understanding of your alarm system's operation and will know when it is working properly and when it isn't. When problems occur, this knowledge will make your opinion more valued by alarm company officials who may prefer to think that there is no problem.

The need for self-testing an alarm system becomes critical as more sophisticated equipment, such as motion detectors, is added. The sensitivity of these highly complex devices is controlled by manual adjustment. Conditions inside your building can upset the balance of a motion detector, causing false alarms or inhibiting their detection of someone in your building. The same problems can result from careless engineering by your alarm company.

You can find out whether your alarm system is functioning by having walk-test lights and/or sonalerts installed on your alarm system. You turn the transmitter onto test mode, walk through your building in the areas the detectors should be covering, and see if the alarm goes off. With a testing light, you may have to ask someone to stand next to the transmitter to see whether the light indicates an alarm. The beeping type makes the job even easier, since you can hear when you have tripped the sensing device.

The internal check of your alarm system goes only as far as the walls of your building. To be sure that any signal activated by your alarm system will actually reach the central station, if you have a non-supervised system, the telephone line connection from your building to the central station should occasionally be checked for its ability to transmit your alarm signal.

The only sure check of the telephone connection is by calling the central station just prior to activating the alarm system and then deliberately setting the alarm system off and having them notify you by phone that the signal is received. Then reset the alarm before leaving.

Never test your alarm system, the alarm company, the police or fire department response without first obtaining approval from all authorities involved.

THE NIGHTMARE ALARM

When the central station is receiving signals from your alarm system caused by something other than your normal traffic or the emergency the alarm is designed to detect, you have a false alarm problem.

When, after repeated service calls, the problem persists with a frequency that is frustrating the alarm company, the police and you, you have a nightmare alarm system.

Your alarm company isn't likely to admit you've got one.

Of course, you can 'bury' this problem by simply ignoring the alarm signals when they seem to fit the false alarm pattern that will develop over time. If you are not paying attention your alarm company may have already used this 'solution' without telling you about it. If this is how your alarm problem is being solved, you must face the fact that you are paying a monthly fee for an alarm system that isn't working.

In this section we will give an outline of procedures for achieving the only solution which really works—finding the causes of the false alarms and eliminating them. This task can be complex or simple, depending on how serious the problem is, of course, but

Fixing it is going to cost money.

also depending on how well you and your alarm company cooperate under circumstances that are trying for both sides.

Over years of experience, we have found that the greatest obstacle to correcting serious alarm problems is the climate of anger and suspicion that rapidly develops between the alarm subscriber and the alarm company when problems first occur.

Face it.

Often, having fallen victim to the glamors of high technology and lacking understanding of the limitations of alarm service companies, the alarm customer holds too high an expectation of what an alarm company can do. Conversely, because of their potential exposure to liability, alarm companies tend to become self-protective, secretive about their work, reluctant to meet their customers eye-to-eye in a frank discussion of a problem.

If you and your alarm company have fallen into this adversary pattern, an otherwise simple engineering or equipment problem can turn into a nightmare.

Sometimes, the first step toward solving the nightmare alarm system is to have a candid, open conversation with your alarm company management— and a willingness to put aside the feelings of frustration and futility that accompany the nightmare alarm system.

Usually, you will find that the solution is one that will take some compromise on both sides. And very often that compromise involves money. A major cause of false alarms is the attempt to rely upon equipment that is out of date or inadequate for the job. Once both parties have reached an understanding of the real source of the problem, they can agree to an equitable solution. But without that understanding, there is no solution.

However, not all alarm problems are the fault of equipment or the alarm company. It often turns out that it is the subscriber himself whose errors are responsible for failures. You must begin the task of solving the nightmare alarm with a willingness to examine your own procedures and attitudes about your alarm system to learn what you could be doing wrong.

Generally, we have found the causes of difficult-to-solve false alarm problems fall into the following categories:

- Antiquated or worn out equipment.

- Manufacturing defects in equipment.

- Damaging or uncontrolled physical environment.

- Subscriber procedural failures.

- Overcomplication of alarms.

- Misapplied, badly engineered equipment.

- Damage by 'expedient' servicemen.

These problems are often interrelated.

Deficiencies in alarm company management can lead to poor installation or rapid deterioration of alarm systems. Constant carelessness on your part can erode the alarm company's concern for your security.

Perhaps the quickest way to locate the source or sources of alarm problems is by a process of elimination, starting first with the most likely causes of false alarms. Below are some questions to help guide this search:

Problem: Antiquated or Worn Out Equipment

How long ago was the alarm installed? If the alarm is more than 15 years old, you can expect to have some problems in both wiring and alarm devices. If it is more than 30 years old, you will certainly be having problems.

Even if you have just paid to have a new alarm system installed, this could still be a factor. To save money, the alarm company could have used wiring from a previous system, resulting in a 'new-old' system and perhaps problems.

If your system is less than 15 years old, you may still have to ask whether it has equipment such as ultrasonic detectors or photoelectric cells that are between 10 and 15 years old. While the wiring of your system may still be intact, the sensing devices themselves may be obsolete. Most likely, the ultrasonic equipment has older electronics that are starting to drift in frequency. There may be no replacement parts on the market and the equipment can't be tuned and held stable.

Solution: Rebuild the whole system or replace all components that are out of date, whether or not they have been proven to be causing problems.

This is probably the most common, but also the most avoided, solution to an alarm problem because it requires a substantial expenditure that both subscriber and alarm company are loath to make.

Complication: It should be no mystery to anyone that once the cost of equipment has been amortized by the monthly service rate over a period of two to two and half years, the monthly charge becomes profit to the alarm company.

If the alarm company completely replaces the equipment and wiring—perhaps spending $1,000 to $1,500 or more in time and equipment—who is going to pay for it? You have already paid an installation fee and are paying a monthly service rate and do not own the equipment. So naturally, you don't feel it is your expense.

The alarm company, on the other hand, does not want to forego five years of profit by installing a new alarm system when it is currently paying for a service force to keep old alarm systems in repair.

If this is your problem, you are going to have to sit down and talk it over with the alarm company, or change companies. But don't jump to the conclusion that your company is a bad one. It simply does not know how to approach you on the problem. Usually you can work out an agreement with the company to reinstall a new alarm system and come out better than if you make a new agreement with another alarm company to put in a new system.

Problem: Manufacturing Defects in Equipment

There are really three problems under this category:

1) There are wide differences in the quality of manufacturing of some of the basic alarm equipment.

2) Sometimes even the best alarm equipment will have a faulty component that is difficult to find because it will intermittently cause false alarms without the component actually failing.

3) Alarm companies must constantly introduce new alarm equipment which always goes through a period of trial by experience before its applications and limitations are fully understood.

This is not just a problem in highly complicated electronic equipment. When you are an alarm company installing thousands of magnetic door switches, and the price ranges from $2 for a marginal brand to $6 for the best, there may be a temptation to sacrifice long-term service for short-term savings. With the $2 door switch, simply slamming a door is likely to cause a failure.

Alarm companies must constantly experiment with new devices and new brand names for potential savings. You have to expect to encounter defective equipment sometimes, but you should be aware of this problem as you compare the bids of different companies.

Solution: The solution to these problems is fairly obvious. The defective piece has to be found and replaced (even if that means replacing every magnetic switch in your premises with a better brand.) If this is the problem, you may have to move your discussions with the alarm company up to a management level because often service personnel are not authorized to make major equipment replacements.

Problem: Physical Environment

If your alarm system is newer than 10 years old and you are still having problems, you should ask whether the physical environment of your building can be highly destructive to alarm equipment. Is the environment very dirty, greasy, damp, exposed to caustic chemical atmosphere or metallic dust from electroplating or manufacturing processes? Is the alarm system exposed to abuse by people hanging coat hangers on wires, bumping into supports that hold photocells, banging into alarm mounts and stanchions? Are any windows cracked underneath foil, or are skylights not aligned properly?

Are there loading doors that no longer seal tightly or can be shifted by the wind? These cause aggravating—and difficult to detect—problems. Sometimes everything seems secure as you leave, but then a strong breeze at 3 a.m. pulls a door ajar enough to momentarily release a switch. Upon your arrival to investigate the alarm, all appears secure.

One of the most important questions: is your building weather tight? An intermittent ground in the alarm system can be caused by a small amount of water seeping through a skylight to wiring, but evaporating before it drips far enough down to be seen.

Solution: Repair the roof! Repair broken windows. Ask your alarm company to send a serviceman to go over your building with you as well as repair or replace any damaged equipment. Take all measures possible to reduce corrosion by moving sensitive equipment away from sources of caustic chemicals. Whenever possible, eliminate the sources of corrosion. Check all doors and windows for snugness. Clean the lenses on all photo cells.

Review with employees the necessity of protecting alarm wiring and devices from casual abuse.

When the alarm needs service, see to it that someone from your company is there to meet the serviceman to let him in quickly, and stay with him to understand your specific causes of service problems.

Problem: Procedural Failures

If your building is fairly clean, if the alarm system is out of reach of employee abuse, and is fairly new, then you should begin to ask yourself if someone could be misusing the alarm. Is there a swing shift or night shift? Do different employees exchange the duties of turning on and off equipment? How do employees view the alarm company? Do the employees understand how the alarm system works?

Could an employee or an automatic timer be switching off an outlet that supplies power to an alarm device such as ultrasonic, microwave, etc.?

Solution: Keep track of the number of keys and passcards to the alarm system, Keep the alarm company updated on the number of people authorized to come and go at night. Review with employees their duties for turning on and off the alarm and for notifying the alarm company of any variation in procedure.

Invite a service supervisor to meet with your employees and go over the potential problems they can help prevent.

Problem: Overcomplication of Alarms

Once the problems listed above have been eliminated as possible causes of false alarm, it is time to begin asking questions about the engineering of your alarm equipment.

Have you overcomplicated your alarm system with non-standard features? Overcomplication frequently begins in the bidding stage when you ask for features that make your alarm system simpler in your mind but are not standard with the alarm company. The alarm salesman, eager to get your business, agrees to try it. The problem occurs when the alarm serviceman comes out to work on the system and encounters something he has rarely seen before. He has no parts or guidelines to service unusual features. A common example of this is the overcomplicated entry/exit system with digital on/off controls at one exit, a key control at another exit.

Solution: To avoid overcomplicating your system, ask the alarm company what it recommends and be sure you are not demanding something beyond its service capabilities. What your alarm company does best will work best.

Problem: Misapplied, Badly Engineered Equipment

Does all alarm equipment and the system itself have adequate ground to a cold water pipe somewhere in the building? Is the area of your business subject to brownouts or intermittent power failures that could be generating false alarms? Are insects, rodents, birds or bats getting into the alarm equipment?

If you have microwave, ultrasonic or infrared equipment in your system, it could be that it is causing false alarms due to improper installation or adjustment. You may be able to detect the problem by asking:

Is all equipment that would be susceptible to radio frequency interference properly shielded? Is the building under the jet flight path of an airport? Is there a train track nearby? Does machinery inside the building emit vibrations or frequencies that the alarm devices are sensitive to? Do the frequencies emitted by telephone bells have any effect on the ultrasonic equipment? Is there a fan that turns on and off automatically in the field of vision of a microwave detector? Is there a heat source, like a pilot light that turns on and off, possibly triggering an infrared detector? Does air conditioning or central heating turn on and off during the night?

Solution: If you suspect there is a specific piece of equipment such as ultrasonic causing the problem, ask the alarm company to connect it to a device called a latching annunciator. When the device causing the problem alarms during the night, the annunciator will lock an indicator light on. The next morning it can be seen that the false alarm orginated in that particular unit.

If the problem is an extreme one, ask the alarm company to add additional transmitters so that each portion of the alarm system—factory, warehouse, office section, ultrasonic unit, photo cells—is on its own alarm transmitter. The central station would know before sending a serviceman what section of the alarm is causing the problem. Once it is isolated, the balance of the alarm system can be used reliably while any alarm eminating from the problem section—for example a failure on a string of ultrasonic heads—can be kept within the alarm company service department until the offending unit or engineering problem is eliminated. This is a built-in advantage offered by some Multiplex systems because they have many alarm zones coming into the central station from each transponder.

Complication: Are you, and is your alarm company, willing to devote several hours in the building on several consecutive nights to observe possible external causes of false alarm? Something as subtle as a fluorescent light that is coincidentally on a wiring circuit with another piece of equipment turned on by a timing device can, without anyone even realizing that it is going on and off, upset an ultrasonic detector. The only way to pin down a cause such as this is to observe it. This is an unusual case, but it points out the difficulties in locating some of these problems.

Problem: Damage by Servicemen

Alarm serviceman often cut into the wire of a perfectly good alarm circuit to test a portion of the circuit. They then twist the wire back together, resulting in a 'twisted splice.' The correct procedure is to then solder and tape the splice to guarantee continuity. All too often, through haste or lack of proper equipment, he will simply twist the wires together by hand and leave. This will leave a connection subject to intermittent failure. This is one of the most common problems and one of the most difficult to find. This is a situation in which a serviceman creates one problem while attempting to solve another.

A related, but far more dangerous problem, can occur when a serviceman deliberately disables a troublesome device. If, after many service calls, your alarm system is no longer causing false alarms, you cannot automatically conclude that your old problems have been solved.

Alarm servicemen often work under the pressure of dozens of service calls per day. Faced with this situation, they have to look for the quickest solution to an alarm problem, not necessarily the one that really solves it. It is not uncommon for an alarm serviceman to solve an ultrasonic problem on a one-night basis by adjusting the equipment sensitivity to so low a level that it would not pick up a tank rolling through your business. Or, equipment suspected of causing problems may simply be cut out of the system. When he does this, the serviceman is supposed to then post the emergency action with the central station supervisor so that, on the following day, service personnel will carry out the needed repairs. Sometimes the report just gets lost in the system and the equipment is never repaired or restored to the alarm circuit. You would not know it because your system would clear to set as you leave each night.

Solution: If you suspect anything in your alarm system has been handled this way, schedule an appointment with an alarm sevice supervisor to go through the entire alarm system with you and ensure that each portion is properly functioning. It is the practice of most alarm companies to leave a service note for you explaining what happened when a man comes in at night to service the alarm system. Your office should have a method to follow these notes up to ensure that proper action has been taken.

Problem: Had Enough?

We have given you a lot of questions to work with. Some of them may help you to solve the problem you are having.

But we hope, above all, to have helped you see the necessity for maintaining close communication with your alarm company to show both that you care about the effectiveness of your alarm system and are willing to compromise, if necessary, to make it work.

We conclude our discussion of alarms on this sober—but we hope not pessimistic—note, that nightmare alarms do occur, but they do not have to occur. As we have attempted to show throughout this book, your decisions make a difference.

Solution: As Nero Wolfe said to Archie Goodwin, 'In the absence of specific instructions, use intelligence guided by experience.'

APPENDIX A

Standards For The Installation, Maintenance And Use Of Central Station Burglar Alarm And Fire Protective Signaling Systems.

The UL regulations comprise many volumes of Standards. This summary of Standards should by no means serve as a final check-list for central station approval. Only a qualified UL representative should attempt that determination.

Rather, this appendix should underscore the high standards of excellence met by all UL Listed central stations and the many possibilities for deficiency in any alarm company that does not have a UL Listed central station.

Perhaps a way of looking at it would be to ask yourself what would happen if something went wrong. Would your alarm company have anticipated the situation and be able to maintain protection of your property? If it is a UL Listed central station, these are some of the things you can take for granted:

General

You certainly want to be confident that your security is in the hands of someone who has experience, professional concern and a personal interest in the success of the work he is doing.

UL attempts to assure you of this, requiring that the central station be controlled and operated by a person, firm or corporation whose principal business is the furnishing and maintaining of a supervised protective signaling service.

There must also be competent and experienced supervisors and operators of mature age present in the central station at all times, 365 days a year. While the number of operators must always be sufficient to handle all contingencies, there must under no circumstances be fewer than one operator plus a guard or runner present.

This is to assure that the central station cannot be unattended (with the sole operator out to lunch or asleep) in the critical first moments after an alarm signal has come in.

The Building

A sophisticated burglar could possibly defeat an alarm system by creating a diversion or assault at the company's central station while his accomplice is hauling away the goods, undetected.

UL therefore requires that the structure housing a central station shall be fire resistant and have 24-hour security against unauthorized entry.

The building shall also be locked at all times and secure from unwanted intrusion. Access shall be limited to authorized personnel.

Power

Power failure, a harmless enough occurrence in the normal urban dwelling, could be disastrous for the central station, which depends for its effectiveness on continuous electrical linkup with its alarm systems. UL has many requirements to prevent interruption of power in the central station.

Most important, there must be two sources of power, a primary and a backup that will switch on automatically in the case of primary failure through a natural disaster, a brownout, etc.

Primary power shall be provided either by a commercial light and power service and an engine-driven generator, or two light and power services supplied on entirely separate networks from independent generating stations.

Many requirements must be met to reduce to a minimum the possibility of interruption of power. The circuit service disconnector, for example, must be installed so as to preclude access to all but authorized employees.

The secondary power supply must be capable of lighting the entire working area and maintaining operation of all alarm equipment continuously for 24 hours.

Secondary power can be provided either by generator or commercial wet cell batteries. In the case of wet cell batteries, adequate charging equipment must be provided and the life of the batteries must be at least twice the charging time.

Wiring

Equally important as the power supply, is the efficient and reliable operation of all electronic circuits within the central station. UL establishes strict standards as to type, gauge and coding for all wiring necessary for the actuation or operation of signaling devices.

Care must be given in the installation of all wiring so that vibrations or jarring cannot cause failures.

There are special requirements for the arrangements of wiring within the central station and for the telephone cable so that a single break, ground, or short will not prevent the transmission of an alarm and that such a break, ground, or short will be indicated automatically at the central station by a trouble signal compelling attention and readily distinguishable from an alarm.

There are also limits on the number of premises that can be attached onto single McCulloh and Multiplex circuits into the central station to prevent overloading.

Signals and Reports

UL listed alarm equipment required in the central station is designed so that the operator cannot miss or ignore a signal. Every alarm must come in three forms—something to see, like a blinking light; something to hear, like a buzzer; something to touch, like a computer printout that the operator can take in his hands.

The time and date of all signals showing a change in status at a protected property must be recorded in a form to expedite prompt operator interpretation.

A written or printed record of all signals received shall be taken and kept for inspection, assuring that the alarm company can be held accountable for the actions of its employees in any occurrence.

Testing

The central station operators must be continually alert to the possibility of equipment failure. UL requires manual tests of all circuits extending from the central station and of all receiving devices in the central station every 12 hours.

In addition, specified modular receiving equipment must be on hand to make an immediate replacement of any defective actuating and transmitting device.

...We hope this brief outline has been useful in establishing the importance of the standards and testing laboratories such as UL in rating the reliability of the company in which you place your trust for the protection of property and lives.

An important point to remember is that while any alarm company may do a good job in most situations, without UL Listing it is hard to be sure its owner has anticipated and prepared for all the situations experience has shown can break down alarm service.

UL approval is not a guarantee of perfection. But it is certainly one of the best available guides to help you discriminate among alarm companies.

APPENDIX B

UL Certification Of Burglar Alarm Systems

The Underwriters Laboratories, Inc. certification of an individual alarm system relates to rating the extent, or thoroughness, of coverage at the protected property, the type of phone connection to the central station and the time taken by alarm company patrolmen to reach the property after an alarm sounds.

As mentioned frequently above, an alarm system is only as good as the central station it is attached to. So, to ensure adequate monitoring of Certificated systems and response to their signals, UL makes additional demands on the staffing, equipment and procedures in the central station for approval of individual alarm systems.

The central station, for example, must have on record a written schedule of the regular opening and closing times at each property covered by a Certificated system and must work out a procedure with the owner to allow variations in the schedule. All openings prior to the usual or scheduled time must be investigated by company guards unless prearranged by special rules.

If, at closing time, trouble exists which cannot be corrected by the subscriber, or if trouble develops upon exit of the subscriber, the operating company must dispatch a repairman after the trouble is detected.

When the alarm company holds keys for the subscriber's property, (and UL provides strict procedures for the handling of keys) every alarm dictates a complete search of the premises and of such adjacent locations as are accessible, together with a subsequent written report to the subscriber on the use of the keys.

When the company does not hold the keys, every alarm dictates a complete exterior search of premises by guards who shall remain at the premises if the subscriber, upon being notified, advises that he will arrive within one hour to allow entry.

At least two alarm company employees capable of acting as guards must be on hand during all closed hours of subscribers' premises.

In addition to records of all alarms and responses, the alarm company must keep records of the actual time of opening and closing of subscribers' premises each day.

Rating of Systems

All UL Certificated alarm systems are rated with a number and a letter. The lowest rating is C-3 and the highest is AA-1.

The number refers to the 'extent' of coverage on the protected property. There are three extents, 1, 2 and 3. Following are the basic requirements for each, as stated in the UL Standards:

Extent 3. This is the lowest 'extent' and can be satisfied with one of four alternatives:

Perimeter protection only—complete[1] protection such as panels, wood dowel screens, etc., or foil and contacts on all accessible[2] windows, doors, transoms, skylights and other openings leading from the premises, **or**

Perimeter and internal channel protection—contacts only on all movable openings leading from the premises and providing one or more invisible rays or channels of radiation (microwave, photocell, infrared, ultrasonic, etc.) equivalent in length to the longest dimension of the enclosed area or areas to detect movement through the channel, **or**

Perimeter and internal volumetric protection—contacts on doors only leading from the protected area or areas; and providing a system of invisible radiation to all sections of the enclosed area or areas so as to detect four-step movement, **or**

Perimeter and sound volumetric protection—contacts only on all movable openings leading from the premises; and providing a sound detection system around the perimeter of the area, with additional microphones located near fixed and movable floor and ceiling openings so as to meet the specifications for Decibel (Db.) ratings.

---

[1]'Complete' refers to devices which employ a double circuit with positive and negative current flow as opposed to foil and contact devices which have only positive current flow and to systems which incorporate end-of-line resistance.

[2]Generally, accessible openings are those 18 feet or less above ground level but in specific cases, clarification should be sought from UL headquarters in Santa Clara.

Extent 2. This 'extent' indicates a strengthened system and also has four alternatives:

Perimeter protection only—complete protection such as panels, wood dowel screens, etc., or foil and contacts on all accessible windows, doors, transoms, skylights and other openings leading from the premises; plus contact devices on all inaccessible[3] movable openings; and also protecting all ceilings and floors not constructed of concrete and all halls, partitions and party walls enclosing the premises, **or**

Perimeter and internal channel protection—contact devices on all inaccessible windows, plus panels, wood dowel screens, etc. or foil and contacts on all accessible windows, doors, transoms, skylights and other openings leading from the premises; and providing a network of invisible beams (radiation devices) to subdivide the floor space of each floor or separate section of the protected area into three approximately equal areas, and more where necessary to provide at least one subdivision per 1,000 square feet of floor space, **or**

Perimeter and internal volumetric protection— supervisory contacts only on all movable openings leading from the premises; and providing a system of invisible radiation to all sections of the enclosed area, so as to detect four-step movement, **or**

Perimeter and sound volumetric protection—contacts only on all movable openings leading from the premises and providing a sound detection system in all sections of the enclosed area so as to meet the specifications for Decibel (Db.) ratings.

Extent 1. This is the strictest 'extent' and may only be used in conjunction with central station alarm companies. It does not offer alternatives:

Completely protecting all windows, doors, transoms, skylights, and other openings leading from the premises, and all ceilings, floors and halls and party and building walls enclosing the premises, except building walls which are exposed to street or public highways, and except that part of any building wall which is at least two stories above the roof of an adjoining building.

This means the building is covered against intrusion even through walls in every part except those out of reach or in public view.

[3]Generally, openings more than 18 feet above ground level.

The **'Grade'** letters in a UL Certification refer to the maximum time required for alarm company patrolmen to reach the premises after an alarm is received and to the type of phone company line connecting the protected premises to the central station. There are three letter grades—C,B and A. The requirements are:

Grade C. The maximum elapsed time from receipt of an alarm or unauthorized opening until patrolmen arrive shall not exceed 30 minutes under normal conditions.

A McCulloh Loop telephone line with a maximum of 15 subscriber transmitters is the minimum acceptable connection for Grade C.

Grade B. The maximum time of response shall not exceed 15 minutes. The minimum line requirement and loading is the same as in Grade C.

Grade A. The maximum time of response shall not exceed 15 minutes from receipt of an alarm.

Grade A Certification requires either a direct line or a Multiplex carrier. If a local bell is connected outside of the premises when a McCulloh Loop is used, the maximum Grade can be 'A' if the guard response time is within 20 minutes.

Grade AA, BB & CC. These are line security systems which are designed specifically for use in districts or on types of premises which, because of location, concentrated values contained, character of merchandise or other factors, may attract a more scientific or highly skilled burglar.

While there appears a considerable difference in the coverage afforded by the lowest Certificated system and the highest, you should keep in mind that UL Certificated alarm systems provide a high level of security and are usually superior to those that do not qualify for Certification.

Blueprint of a Protection System and Index

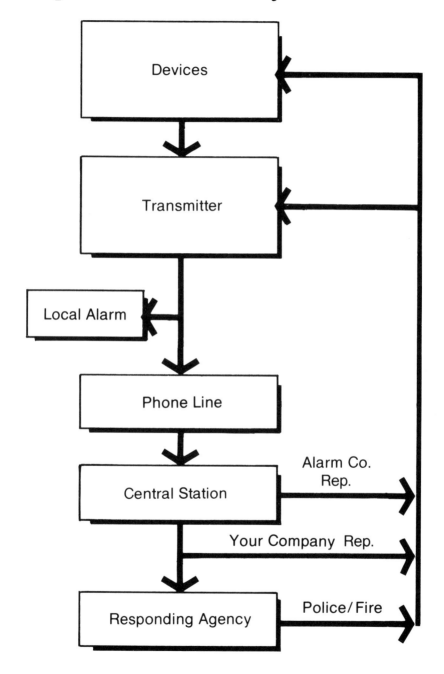

Alarm Sensing Devices and Circuits

On Premises

Burglary & Holdup

Contact Switches . . . *all types* 45-6, *uses* 63, *outdoor* 65, *home* 125-6, 127, *safe* 62, *high-security* 43-4, 89

Foil . . . 46, *uses* 60, *strip* 47, *home* 125

Alarm Screens, Wire Lacing & Panels . . . 48, 60-1

Window Shock Sensor . . . 57, *home* 125

Vibration Detectors . . . 57, *outdoor* 65

Sound . . . 58

Light Beams . . . 49, *uses* 60-1, *home* 126

Infrared Sensors . . . 55-6, *uses* 60-1, *outdoor* 63, home 126

Ultrasonic Sensors . . . 50, 51-53, *uses* 60, 63, *home* 126

Microwave Sensors . . . 50, 53-5, *uses* 60-1, 63, *outdoor* 64, *home* 126

Radar Sensors . . . 50, 53

Capacitance Alarm . . . 56

Vault Alarm . . . 58, 61-2, *UL* 88-9

Holdup Devices . . . *all types* 66-8

Special Devices

Cameras & CCTV . . . 90, 91

Card Access Systems . . . 91

Emergency Notification . . . 92

Proprietary Alarm Systems . . . 91-2

Home

Home Control Center . . . 121, 125-6

Interior Protection . . . 124-5

Panic Buttons . . . 121, 127

Outside Gates . . . 127

Swimming Pools . . . 127

Medical Alert . . . 127

Self-Contained Devices . . . 54, 104, 134-5

Fire

Manual-Pull Stations . . . 103, 104-5

Fire Detectors . . . *all types* 103, 106-8, *home* 122-3, 134

Water Sprinkler Monitors . . . *all types* 103, 109-11

Valve Switches . . . *all types* 110-11

Insurance . . . 100-01, 112-13

Industrial Hazards . . . 97, 113-14

General

Pre-Wiring . . . 119, 132

Normally Closed vs. Normally Open . . . 43

Active vs. Passive . . . 44, *home* 123

Tampering . . . 63, 78, 125

End-of-Line Resistance . . . 44, 76,78

Standby Power . . . 76, 80-1, *fire* 99, *home* 123, 124

Zoning . . . 42, 76, *fire* 113, *uses* 145-154, *home* 124

On/Off Stations . . . *all types* 29, 124-25

Time-Delay . . . 82, 124, 130

Self-Testing . . . 163-5

Voice Simulators . . . 131

Wireless Alarm Systems . . . 133, *fire* 134

Security Bars or Gates . . . 6

Equipment Bolt-Down Devices . . . 33

Judge Anderson created by John Wagner and Brian Bolland

Anderson Psi Division: Shamballa
created by **Alan Grant** and **Arthur Ranson**

Steve Potter
Lettering

Richard Burton
Commissioning Editor

Steve Edgell
Graphic Novel Editor

Steve Cook
Design

ISBN 1 85386 247 9

Copyright 1991 2000 AD Books, a division of Fleetway Publications, Greater London House, Hampstead Road, London NW1 7QQ, a member of Maxwell Consumer Publishing & Communications Ltd. UK Distribution and Export by MacDonald & Co (Publishers) Ltd, Tel (071) 377 4600. USA Representation and Marketing: SQP Inc, 137 Pacific Avenue, Beechwood, NJ 08722, USA.

Steve MacManus
Managing Editor

Jon Davidge
Managing Director

Sal Quartuccio
Bob Keenan
US Sales Directors

Anderson Psi Division: Shamballa first appeared in **2000 AD** Progs 700-711, between 13 October and 29 December 1990

Origination by Hoistmuir, Birmingham
Printed and bound in Spain by Cronion SA, Barcelona
First edition July 1991

1 3 5 7 9 10 8 6 4 2

ANDERSON PSI DIVISION

SHAMBALLA

ALAN GRANT

ARTHUR RANSON

2000 AD BOOKS

THEY USED TO SAY WHEN SOMEONE DIED, A NEW STAR SHONE IN HEAVEN.

THAT ONE'S COREY'S.

WITH TELEPATHS, FRIENDSHIP RUNS DEEP. WE KNEW EACH OTHER FROM THE INSIDE— SHARED EACH OTHER'S HOPES AND FEARS AND HIGHS AND LOWS—

HELPED EACH OTHER MAKE IT THROUGH THE MADNESS OF MEGA-CITY LIFE.

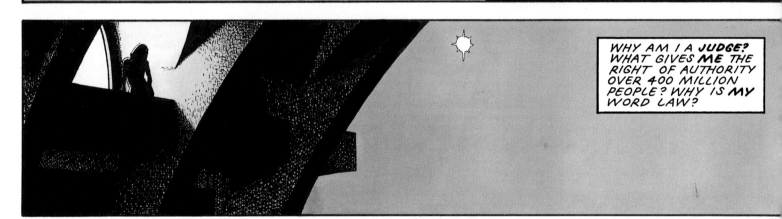

WHY AM I A *JUDGE?* WHAT GIVES *ME* THE RIGHT OF AUTHORITY OVER 400 MILLION PEOPLE? WHY IS *MY* WORD LAW?

NOW SHE'S GONE, AND A PART OF ME IS, TOO, AND SOMEHOW IT'S GOT A WHOLE LOT HARDER TO GIVE MYSELF GLIB ANSWERS WHEN I ASK MYSELF "WHY?"

WHY DO NICE THINGS NEVER HAPPEN?

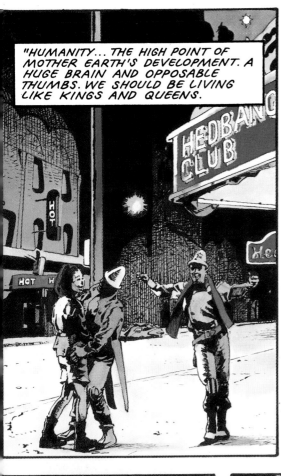

"HUMANITY... THE HIGH POINT OF MOTHER EARTH'S DEVELOPMENT. A HUGE BRAIN AND OPPOSABLE THUMBS. WE SHOULD BE LIVING LIKE KINGS AND QUEENS.

"INSTEAD, WE'RE JUST WELL-DRESSED MONKEYS—"

JUDGE = NAZI

HEY! NICE PUSSY—! C'MON, BOY! HOWJA LIKE A LITTLE SPRAY JOB..?

AAH-AAH-AHH-

YOU SHAY "CAT", AL? ISH 'AT A CAT? LET'SH HAVE SHOME—

...fun!

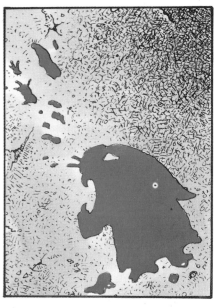

REPORT: to EAST-MEG 2 DIREKTORAT from PsiKop Amisov
DATE: 17.1.12

I was coming off duty after a standard night patrol in the company of Judge Yenko.

While returning to the Sektor House I experienced a vivid psionic impression—

It was the image of some creature, like a bear yet not, with slavering jaws and green glaring eyes—

I barely had time to react before the beast was on us—

YENKO! LOOK OUT—!

In the *seconds* it took me to recover and draw my weapon, Yenko was brutally slain--

Despite my fire, the beast seemed untroubled. It turned its attention to me, and I swear that before it disappeared--

It smiled at me.

SUBJECT: AMISOV

Interogation : Confirms
Lie Detector : Confirms
Deep Probe: Confirms

SWEET GRUD!

WHO THE HELL COULD DO SOMETHING LIKE THIS..?

NOT WHO. *WHAT.*

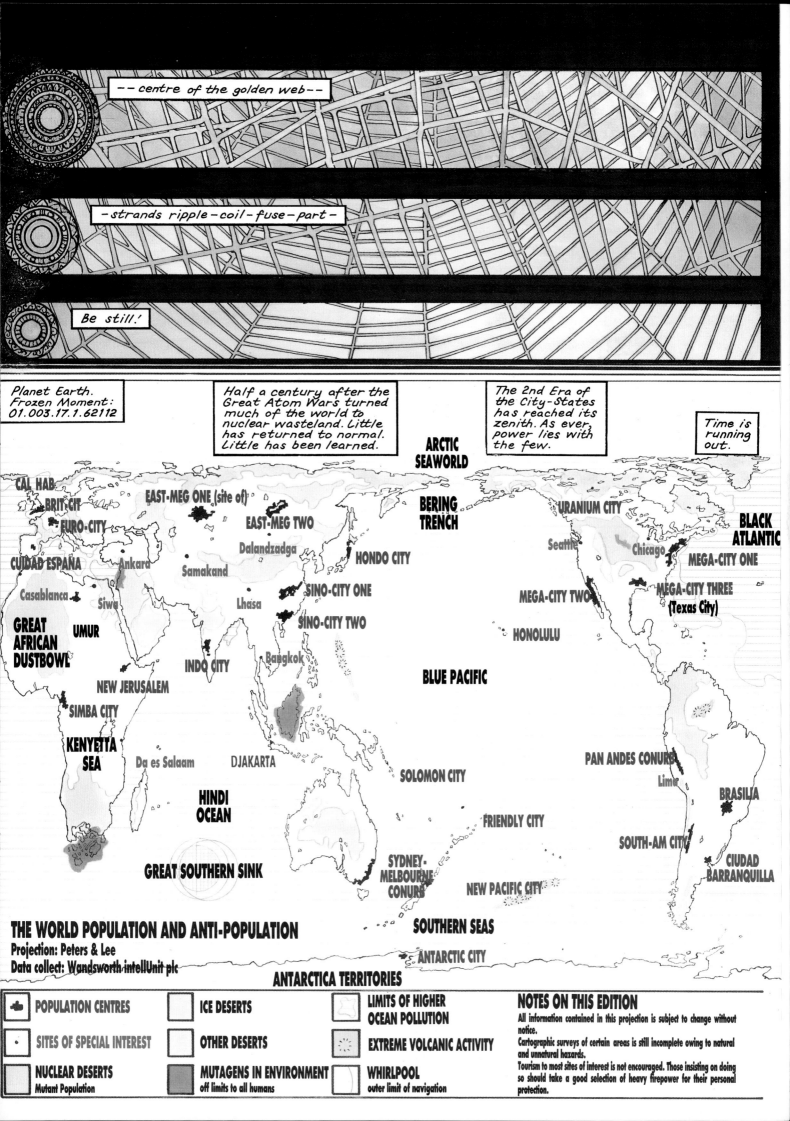

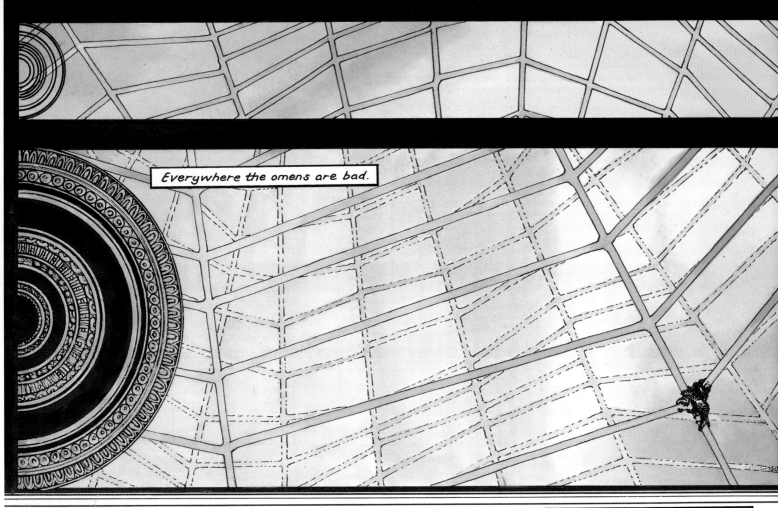

Everywhere the omens are bad.

Brit-Cit: 419 sightings of the Black Dog in one week. Thanatonia up a thousand percent.

CuidadEspana: 73 stigmatics spontaneously mimic the Sacred Wounds. Religious fervour.

Oz: Suburbia terrorised by phantom aboriginal Bunyip. The deaths are real.

Djakarta: The Manticore has surfaced. Born again Cannibals hold high government office.

SouthAm: Massed ghosts of The Disappeared Ones witnessed by millions.

Hondo-Cit: Oni demons seen walking on Mount Fuji. The cherry harvest failed. Suicides trebled.

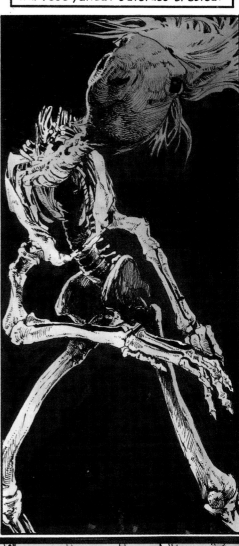

--centre of the golden web--

The omens are bad. The strands sever.

Time is running out.

-- strands reach out, questing--

--small lights in a great darkness--

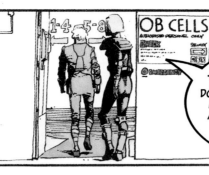

-- pinpricks in infinity --
DEPARTMENT OF FORTEAN EVENTS

...FOURTH ATTACK IN THREE NIGHTS, DOC! THE SAME STORY— MINCED CITIZENS, AN' PAWPRINTS THAT GO *NOWHERE!*

YOU HAD A *PSI-FLASH* BEFORE THE LATEST, I BELIEVE?

YEAH. GIANT URBAN CAT— STANDARD FORTEAN IMAGE. WITH GREEN GLARING EYES.

THEN SOME GUY— AN' A SCENE STRAIGHT OUT OF *HELL.* COULDN'T TELL *WHAT* THE DROKK IT MEANT!

MEANING... THE *BANE* OF *ALL* OUR LIVES !

UHH... SORRY. COME AGAIN?

NEVER MIND. HERE— I WANT YOU TO SEE THESE...

LEVITATING NUN. INVOLUNTARY. WE CAN'T BRING HER DOWN.

TRANQUIL- LISED— STOP HER BANGING HER HEAD ON THE CEILING!

I'VE HEARD OF GETTIN' HIGH ON RELIGION, BUT...

LITTLE SISTER OF GROMINK

MEET THE DARWINS. ALL FOUR OF THEM GREW TAILS OVER- NIGHT.

AARON CHAGAL. SAYS HE'S POSSESSED BY THE HELL-DEMON *BARBAARIK.* NOT SO UNUSUAL— EXCEPT WE HAVE TWENTY-THREE OTHERS CLAIMING THE SAME!

YOU THINK THIS IS AN EXTENSION OF THE *'TRIAD'* CAPER? *ORLOK* UP TO HIS DIRTY TRICKS?

NO. I HAD A CALL FROM MY COUNTERPART IN EAST-MEG TWO, *PROFESSOR LYCHENKO.*

THERE'S A NEW SPIRIT OF GLASNOST AROUND. THEIR JUDGES HAVE *DISOWNED* ORLOK — HE AND HIS FOLLOWERS HAVE BEEN RE-DEFINED AS 'MISGUIDED TERRORISTS'.

BESIDES, THE EAST-MEGS ARE HAVING THEIR OWN BOOM IN FORTEAN EVENTS — LIKE HALF THE *REST* OF THE WORLD!

FACT IS, THAT'S WHY LYCHENKO CALLED. HINTS THAT *SHE* KNOWS WHAT THE *CAUSE* OF ALL THIS INSANITY MIGHT BE. WANTS ME TO GO OVER THERE.

DON'T TOUCH! METEORITE, PRE-COLUMBAN. UNKNOWN SUBSTANCE.

OH. AND ARE YOU GOING? EAST-MEG, I MEAN?

YES. OF COURSE, CHIEF JUDGE DOESN'T TRUST ME ALONE. INSISTS I TAKE A BODYGUARD.

KKRKK

KRAKK

BRUUPPP!

YOU *SURE* IT'S WISE LETTING THOSE TWO CLOWNS GO OFF ON THEIR OWN?

DOC RICKARD'S A GOOD PSI — BUT HE'S WAY PAST HIS BEST. AND *ANDERSON* HARDLY SEEMS THE LOGICAL CHOICE FOR *BODY-GUARD* DUTY — SHE'S TAKEN IT HARD SINCE COREY DIED.

WHAT IF ALL THIS GLASNOST TALK IS JUST SO MUCH *BULL?* WHAT IF THE SOVS *ARE* UP TO SOMETHING? WE'D BE HANDING THEM TWO TOP TELEPATHS ON A PLATE — !

DREDD, WHATEVER'S GOING ON IN THE WORLD, IT'S WAY BEYOND THE PETTY RIVALRIES BETWEEN MEGA-CITY AND EAST-MEG TWO.

IF THIS LYCHENKO THINKS *SHE* KNOWS SOMETHING ABOUT IT, AND HE THINKS RICKARD CAN HELP — SO BE IT. AT LEAST IT'S A POSITIVE ACTION.

NO OFFENCE, CHIEF JUDGE, BUT IT COULD TURN OUT POSITIVELY *STUPID!*

YOU'RE GOING TO HAVE TO TRUST ME ON THIS ONE, DREDD. IT'LL BE ALL RIGHT.

WHAT MAKES YOU SO SURE?

CALL IT A GUT FEELING...

THREE MILES HIGH. PEOPLE SO SMALL AND INSIGNIFICANT YOU CAN'T EVEN SEE THEM.

I COULD USE A CHANGE OF SCENE. SOMETHING TO TAKE MY MIND OFF THINGS.

SOMETHING TO SHUT THE QUESTIONS OUT—

THERE—! *LOOK!* SEE IT?

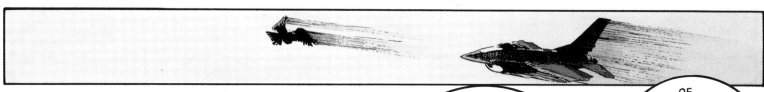

WHAT IN THE NAME OF THE GREEN GRUD IS *THAT?*

THE BLACK EAGLE...

ONE OLD LEGEND HAS IT THE BIRD APPEARS WHEN THE PEOPLE OF THE EAGLE *NEED* IT.

PEOPLE OF THE EA— OH. I SEE. *US.*

OF COURSE, A SECOND LEGEND HOLDS THAT ITS APPEARANCE HERALDS THE *END* OF THE *WORLD!*

YOU REALISE OF COURSE, COMRADE LYCHENKO, THAT THIS COULD BE A MEGA-CITY *TRICK?*

THE DIKTATORAT EVIDENTLY THINK SO — ELSE THEY WOULD NOT HAVE GIVEN US OUR ...MINDERS.

PAH! THE DIKTATORAT LIVE IN THE PAST, *AMISOV.* THEY HAVE PLAYED AT WARS AND SPIES AND SOLDIERS FOR TOO LONG TO BE ABLE TO SEE THE *REAL* DANGER, EVEN WHEN IT SNARLS IN THEIR FACES!

BE CAREFUL WHAT YOU SAY, PROFESSOR. THESE TWO WILL REPORT BACK EVERY WORD!

PAH! LET THEM! EVENTS WILL PROVE ME RIGHT IN THE END.

HOW CAN YOU BE SO SURE..?

I HAVE WHAT OUR VISITORS WOULD CALL A... FEELING IN THE LOWER PART OF MY ALIMENTARY CANAL!

WHAT I AM ABOUT TO TELL YOU IS FANTASTIC— UNBELIEVABLE, EVEN. BUT THEN, WHO WOULD BELIEVE IN *GHOSTS? MANTICORES?* GIANT URBAN CATS? FLESH-EATING *YETI?*

THE WHOLE WORLD IS AFFECTED BY THIS MASSIVE SURGE IN PSYCHIC PHENOMENA. EVERY DAY THE REPORTS ARE OF FRESH SIGHTINGS— NEW DEATHS—MORE MENTAL BREAKDOWNS!

NO JUDGE FORCE IN ANY CITY HAS SHOWN THE SLIGHTEST ABILITY TO CONTROL—OR EVEN *EXPLAIN*— IT!

WITH ALL DEFERENCE, PROFESSOR— WE ALREADY KNOW THIS.

PAH! THE IMPATIENCE OF YOUTH!

VERY WELL. LISTEN AND HEAR SOMETHING YOU *DON'T* KNOW—

THE TERRIBLE *LEGEND* OF AGHARTI... *THE WORLD BELOW!*

"60,000 YEARS AGO, *GIANTS* WALKED THE EARTH.

"THEY WERE IN THEIR HEYDAY THEN, AND THE OTHER TRIBES WERE NO MATCH FOR THEIR SHEER SIZE AND SAVAGERY.

"ACCORDING TO LEGEND, A SHAMAN—*TCHUD*, THEY CALL HIM IN PARTS OF THE *SOV-BLOK*—SURVIVED WITH A FEW OF HIS TRIBE.

"TCHUD SPOKE WITH THE EARTH-MOTHER, AND ASKED HER WHERE HIS PEOPLE COULD GO TO BE SAFE—

"AND THE EARTH OPENED HERSELF UP TO THEM."

TCHUD'S TRIBE FOLLOWED THE CAVES DOWN, DEEP INTO THE BOWELS OF THE EARTH...

NICE STORY, PROF. I DON'T SEE WHAT BEARING IT HAS ON *OUR* PROBLEMS, THOUGH.

PLEASE. LET ME FINISH.

"WHEN THEIR TORCHES FAILED, THEY FOUND CAVES THAT HAD THEIR OWN ILLUMINATION— LUMINESCENCE FROM THE ROCKS.

"FUNGI AND MOSSES GROWING IN THE EERIE GREEN LIGHT PROVED HIGHLY NUTRITIOUS, AND PROVIDED ALL THEIR DIETARY NEEDS.

"AND THEN THEY FOUND *THE CITY.*

"THEY HAD NO WAY OF KNOWING HOW OLD IT WAS. WHO-EVER — OR WHATEVER — HAD BUILT IT, EVEN THEIR MEMORY WAS LONG FORGOTTEN.

"THE SILENCE OF MILLIENNA LAY ON IT LIKE A SHROUD.

"HERE THEY SETTLED, WITH TCHUD AS THEIR LEADER, RESOLVED TO NEVERMORE SEEK OUT THE DANGERS OF THE SURFACE WORLD.

"AND SO A MIGHTY CIVILISATION WAS BORN—AND FLOURISHED."

SO WHAT YOU SAYING, PROF? THAT THEY'RE STILL *DOWN* THERE— THAT THEY'VE NEVER BEEN *FOUND?*

ENTRANCES TO THE SUBTERRANEAN WORLD HAVE BEEN FOUND MANY TIMES.

YES, I CHECKED THE RECORDS AFTER YOUR CALL. MEXICO—PERU —BAVARIA—TIBET...DOZENS OF ANECDOTAL CASES. ALL SAYING THE SAME THING: AGHARTI *DOES* EXIST!

A MASSIVE UNDERGROUND KINGDOM THAT SPANS THE ENTIRE GLOBE! BUT ITS STORY DOES NOT END THERE—

"FOR IT IS SAID THAT, IN THEIR STRANGE NEW WORLD, TCHUD'S TRIBE DISCOVERED THE *FORBIDDEN POWER.* AT FIRST THEY USED IT ONLY FOR GOOD, TO BUILD THEIR UTOPIA...

"BUT THERE WERE THOSE WHO SUCCUMBED TO GREED AND LUST FOR POWER. THE *DEROS,* THEY CALLED THEMSELVES—

"PERVERSE THROWBACKS TO MORE SAVAGE DAYS, THEY *DELIGHTED* IN DEATH AND DESTRUCTION. OUTLAWED, THEY TOOK TO THE DEEP TUNNELS—

"SCHEMING, PLOTTING, WAITING FOR THE DAY WHEN THEIR STRENGTH AND NUMBERS WOULD ALLOW THEM TO *WIN* AGHARTI...

"AND AFTER IT, THE CONQUEST OF THE WORLD ABOVE!"

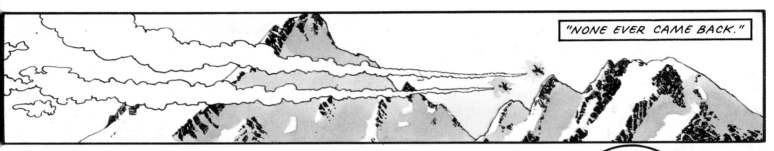

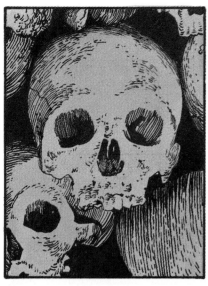

I FEEL LIKE I'VE KNOWN HIM FOREVER...

I FEEL LIKE I'VE KNOWN HER FOREVER...

CASSANDRA—

MIKHAEL—

— ugly things —

— cancer in the golden web —

— spreading —

PLANET EARTH FROZEN MOMENT: 09.080.18.1.62113

A PAN-AFRICAN SUPERTANKER SET COURSE DIRECTLY FOR THE **SOUTHERN SINK**. IT IS THE MAELSTROM'S 666th VICTIM.

IN BRIT-CIT A DESCENDANT OF DORIS STOKES HAS FOUND FABLED **EXCALIBUR** AND PRONOUNCED THE RETURN OF **AVALON**.

THE PACIFIC VOLCANO **NAI-NO-KAMI** SPEWED FORTH A MILE-LONG PLUME OF POISON DUST IN THE SHAPE OF **K'UN**, THE CHINESE IDEOGRAM FOR **OPPRESSION**.

IN THE RUINS OF DELHI, A RENEGADE CULT HAS ANNOUNCED THE START OF THE NEW **TOWER OF BABEL** — TO BE BUILT ENTIRELY OF VOLUNTEERS' CORPSES.

THE SPIRITS OF FOUR ROCK GIANTS PERFORMED AN 8-HOUR VERSION OF "A HARD RAIN'S GONNA FALL" ON A PARIS BOMBSITE.

IN THE VATICAN, THE **TURIN SHROUD** IS WEEPING TEARS OF BLOOD.

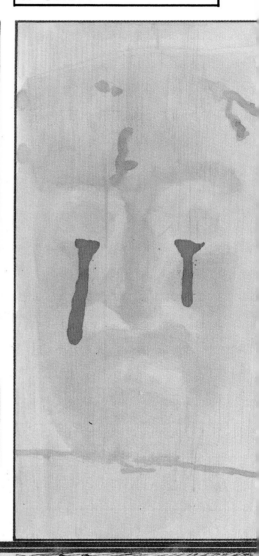

— the omens are screaming —

— the web is split —

— can the centre hold? —

WE DON'T NEED TO SPEAK.

MY THOUGHTS ARE HIS THOUGHTS ARE MY THOUGHTS.

--THREE OF THEM--

--LIKE THE ONES WE SAW IN OUR VISION--

--REEK OF EVIL--

--GOING FOR THE CONTROL POD--

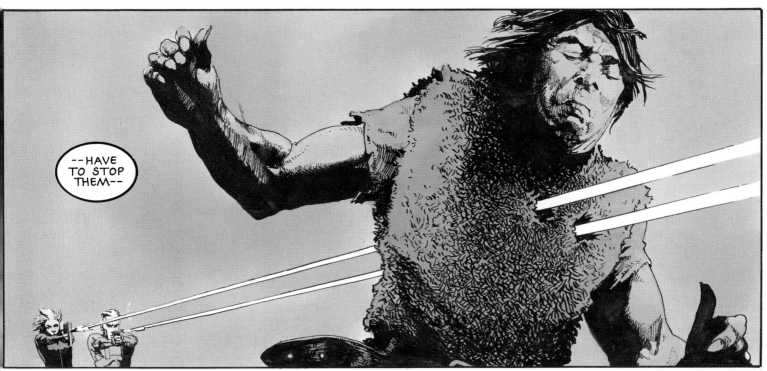

--HAVE TO STOP THEM--

--THE
SENSORS--

--SPEEDING
UP--

--GOING
OUT OF
CONTROL--

CASSANDRA!

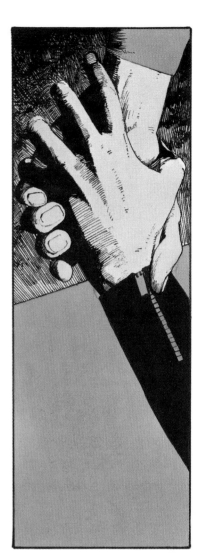

--OH
MY
GRUD!--

20

THEY WERE LIKE MUTANTS. BIG. HAD GREEN EYES.

HOW DO YOU FEEL ABOUT OUR LITTLE EXPEDITION NOW, ANDERSON?

I'M, UH... MORE OPEN-MINDED, SIR. THOUGH I GUESS THE JAUNT ENDS HERE!

OH, I THINK NOT. SOMEONE—OR SOMETHING—OBVIOUSLY DOESN'T WANT US TO CONTINUE.

BUT LIKE ALL GOOD PSIS, WE COME PREPARED...

WE WOULD HAVE HAD TO LEAVE THE TRAIN SOON, ANYWAY.

HOV-KITS! I HAVEN'T USED ONE OF THEM IN YEARS!

AND YOU WILL NOT BE USING ONE TONIGHT!

THAT EQUIPMENT WAS NOT DECLARED TO THE DEPARTMENT.

NO-ONE LEAVES UNTIL WE RECEIVE FURTHER INSTRUCTIONS FROM THE DIKTATORAT!

RUBBISH!

LOOK—

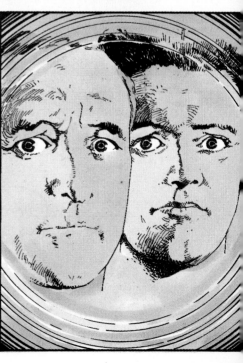

FUNNILY ENOUGH, NO.

THANK GRUD THE SUITS COME INSULATED!

THREE HOURS TO CLIMB TEN THOUSAND METRES. STOP EVERY TWENTY MINUTES TO ACCLIMATISE.

OXYGEN FROM THE CANISTERS LIKE NECTAR TO PAINED WHEEZING LUNGS.

THE MOUNTAINS DWARF US. OUR FRAIL MACHINES CARRY US UP, LIKE GNATS STRIKING OUT FOR HEAVEN...

ANTS ON THE ROOF OF THE WORLD.

FROZEN— ETERNAL— AWESOME. THEY USED TO CALL IT TIBET.

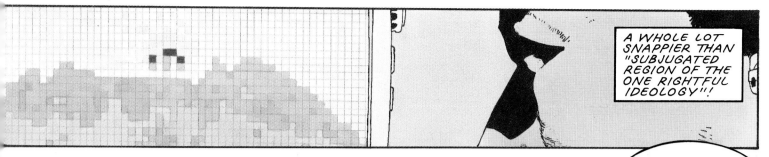

A WHOLE LOT SNAPPIER THAN "SUBJUGATED REGION OF THE ONE RIGHTFUL IDEOLOGY"!

热烈欢迎各国朋友

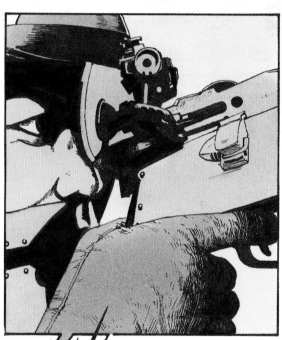

THEY'RE WAITIN'!

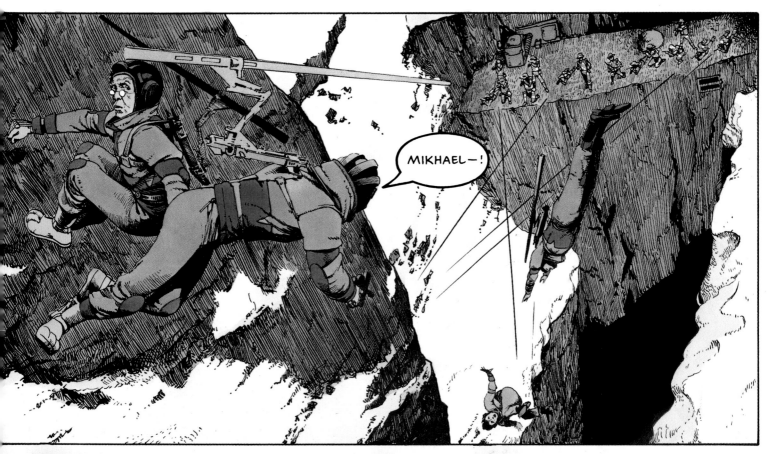

CASS... DON'T BROOD. YOU **HAD** TO KILL THEM.

I KNOW. BUT I DON'T HAVE TO LIKE IT!

I'VE KILLED DOZENS IN MY TIME. WOUNDED HUNDREDS. I'VE BROKEN LIMBS, AND SQUEEZED MEN'S MINDS UNTIL THEY SCREAMED.

"PEOPLE WITH SPECIAL GIFTS SHOULDN'T DO UGLY THINGS." IS IT UGLY TO KILL SOMEBODY WHO'S TRYING TO KILL YOU?

DAMN YOU, COREY! YOU WERE MY **FRIEND**. WHY DID YOU HAVE TO MAKE IT SO **HARD?**

WHAT THE HELL AM I **DOING** HERE?

FIVE MILES HIGH, WITH TWO OLD KOOKS AND A GUY I'D'VE **SHOT** AS THE **ENEMY** NOT A MONTH BACK—

IN THE DARK—

CHASIN' AN OLD LADY'S HUNCH THAT'LL TURN OUT TO BE A TOTAL **CROCK**. CLIMBIN' MOUNTAINS WHILE THE WORLD BURNS.

SOMETHING'S HAPPENING..!

IT... IT'S FADING AWAY—!

IT'S JUST A RUIN!

IT WAS AN *ASTRAL PROJECTION*—AN IMAGE OF HOW GLORIOUS *SHAMBALLA* ONCE WAS.

BUT NOW... WE HAVE COME ALL THIS WAY FOR... *THAT?*

THINK, MIKHAEL! *SOMEONE* — OR SOME*THING* — MUST HAVE *PROJECTED* THE IMAGE!

AND THAT MAGNITUDE OF POWER COULD *EASILY* HAVE CAUSED THE PSYCHIC DISTURBANCES WE ARE HERE TO INVESTIGATE!

I HAVE TO SAY I'VE GOT A BAD FEELING ABOUT THIS.

BUT WHAT THE HELL — WE'RE HERE NOW!

INSIDE, IT FEELS LIKE A MAUSOLEUM—

LIKE, ONCE, SOMETHING BEAUTIFUL LIVED HERE, SOMETHING THAT WORSHIPPED GODS WE NEVER HEARD OF, LIVED LIVES WE CAN'T EVEN DREAM ABOUT.

A SOUL-DEEP SADNESS PERMEATES THE STONES. HIDDEN IN THE SILENCE, ECHOES OF DEATH-SCREAMS.

AND SOMETHING ELSE—

DEROS!

DEEP, BUT CLEAN. I'LL PUT ON A MED-PAK, SPEED UP THE HEALING.

HURRY IT UP! THERE ARE MORE COMING!

TOO MANY TO FIGHT! THIS WAY—!

GRUD! MORE OF THEM!

ZAAK

SHRAAAK!

UP HERE! MAYBE IF WE CAN GET ABOVE THEM...!

IT'S NO GOOD! THEY'RE WAITING!

WE'RE TRAPPED!

HOLD YOUR FIRE!

THEY'VE STOPPED ATTACKING. THEY'RE JUST... SITTING THERE.

AS IF... THEY ARE *WAITING*.

LOOKS LIKE WE GOT A BREATHER, PROF. WHAT YOU FOUND?

OUR WAY OUT, PERHAPS. ALTHOUGH—

THESE ARE *DEATH SYMBOLS*. WHATEVER THE PASSAGE WAS USED FOR, *PSYCHICALLY* IT'S LEFT ITS *MARK*.

IT FEELS... *HORRIBLE* IN THERE.

AIN'T EXACTLY WONDERFUL OUT HERE, EITHER, PROF!

THERE'S *HUNDREDS* OF THEM—AND OUR GUNS ARE LOW ON CHARGE.

SO UNLESS YOU'RE PLANNIN' SOME KIND OF DIPLOMATIC *MIRACLE*—

I SAY WE AT LEAST TAKE A PEEK!

INTERESTING... THE SAME SORT OF LUMINESCENCE MENTIONED IN THE *TCHUD* LEGEND!

WHAT'S THAT SMELL..?

OH *GRUD*..!

DOZENS OF THEM!

SOME OF THEM LOOK AS IF THEY DIED YEARS AGO... YET OTHERS MIGHT HAVE PERISHED ONLY *YESTERDAY*!

THEN SUDDENLY—

I'M IN HELL.

EVERY HORROR—EVERY MONSTROSITY
I'VE EVER SEEN COMES BUBBLING
AND SEETHING FROM THE DARK PITS
OF MY MIND...

FILLING ME—SCRAPING AT THE INSIDE OF
MY SKULL... SQUEEZING MY BRAIN WITH
ROTTING, LONG-DEAD FINGERS...

КассаНДра!

КассаНДра!

CASSANDRA! SNAP OUT OF IT..!

WH-WHAT HAPPENED..?

WHERE ARE—?

END OF THE ROAD, I'M AFRAID. THERE'S NO OTHER WAY OUT OF HERE!

AS FOR WHAT— WE EVIDENTLY TRIGGERED SOME SORT OF PSI-TRAP. IT SENT US ALL ON A VERY BAD TRIP.

THOUGH IT SEEMED TO AFFECT YOU WORST.

PSI-TRAP? BUT WHY WOULD ANYONE—

THE BEST WE CAN GUESS AT...

IS TO PROTECT HIM!

IS HE... DEAD?

ABOUT A THOUSAND YEARS, BY THE LOOK OF HIM!

BUT NO. SOMEHOW— INCREDIBLY!— THERE IS STILL A SPARK OF LIFE IN THERE!

"WE'VE COME ALL THIS WAY—TO THIS RUIN ON THE TOP OF TIBET—TO FIND OUT WHAT THE HELL'S **HAPPENING** IN THE WORLD... AND **HE IS IT**?

"BUT WHY, FOR GRUD'S SAKE? **WHY**?

"THERE'S ONLY ONE WAY TO FIND OUT.

"SOMEONE'S GOING TO HAVE TO GO IN THERE AND **SEE**!"

SOMEONE..?

OH, I GET IT. **ME!**

I WOULD DO IT, CASSANDRA. BUT... THIS BITE. I DO NOT FEEL SO GOOD.

NO PROBLEM, PAL.

I'VE **ALWAYS** WANTED TO DO THIS!

HA HA!

I EXPECT DARKNESS AND MAGGOTS AND CORPSE THOUGHTS—

BUT LIGHT EXPLODES IN A GREAT GOLDEN BALL—

TEN MINUTES— AND SHE HASN'T MOVED OR SPOKEN. REMARK- ABLE!

DO YOU THINK SHE IS...ALL RIGHT?

I HOPE SO. SHE'S THE BEST WE HAVE.

PERHAPS WE SHOULD ASK—

NO!

GOOD GRUD, MAN! DON'T YOU KNOW THE DANGERS OF INTERRUPTING A TRANCE AS DEEP AS THAT? SHE COULD *LOSE* HER *MIND* IN THERE!

OF COURSE. I KNOW. I...I'M SORRY.

I'M JUST NOT THINKING STRAIGHT. IT'S THIS DAMN BITE...

IT WON'T STOP THROBBING..!

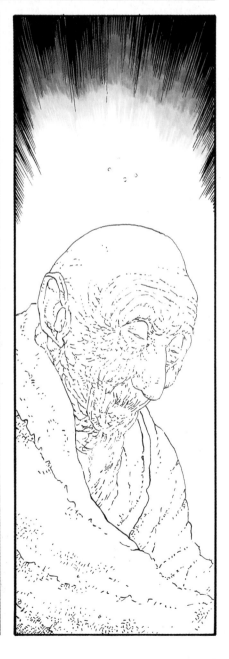

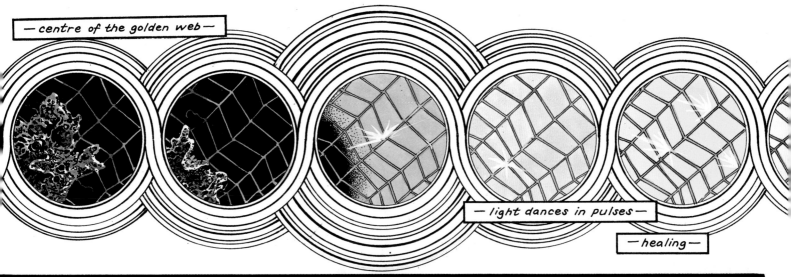

— centre of the golden web —

— light dances in pulses —

— healing —

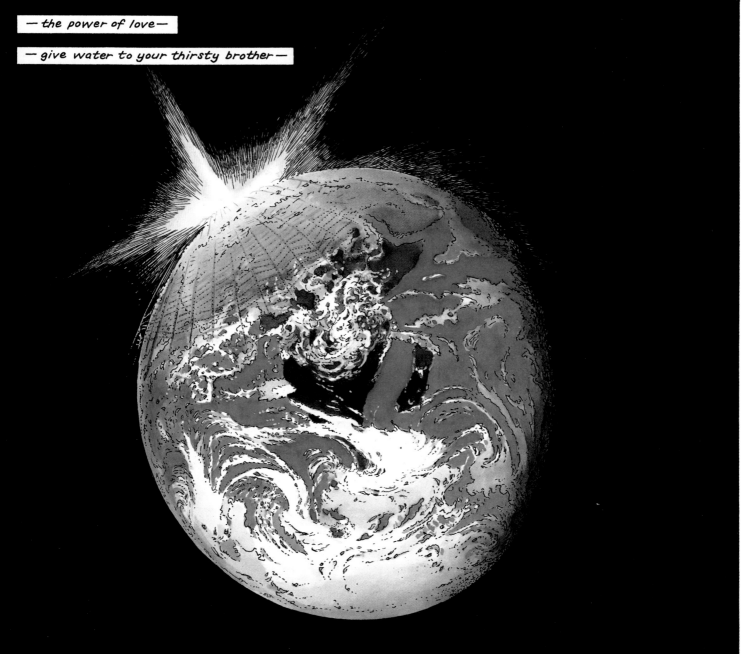

— the power of love —

— give water to your thirsty brother —

— a new star rises in the East —

— sing !

I DON'T BELIEVE ANY OF IT.

I DON'T BELIEVE THERE'S AN ANCIENT MAN SITS ON THE ROOF OF THE WORLD AND KEEPS THE WHOLE CABOODLE ROLLING.

I DON'T BELIEVE THERE'S A RACE OF MALIGNANT SUBTERRANEANS JUST WAITING TO INVADE US.

I DON'T BELIEVE IN SACRIFICE.

I DON'T BELIEVE THAT WHEN A LOVED ONE DIES, A NEW STAR SHINES IN THE SKY.